YOUR DAILY MATH

YOUR DAILY MATH

NUMBER PUZZLES
AND PROBLEMS
TO KEEP YOU SHARP

LAURA LAING

FALL RIVER PRESS

FALL RIVER PRESS

New York

An Imprint of Sterling Publishing
1166 Avenue of the Americas
New York, NY 10 36

ISBN 978-1-4351-6203-7

Distributed in Canada by Sterling Publishing Co., Inc.
c/o Canadian Manda Group, 664 Annette Street
Toronto, Ontario, Canada M6S 2C8
Distributed in the United Kingdom by GMC Distribution Services
Castle Place, 166 High Street, Lewes, East Sussex, England BN7 1XU
Distributed in Australia by Capricorn Link (Australia) Pty. Ltd.
P.O. Box 704, Windsor, NSW 2756, Australia

For information about custom editions, special sales, and premium
and corporate purchases, please contact Sterling Special Sales
at 800-805-5489 or specialsales@sterlingpublishing.com.

Manufactured in China

2 4 6 8 10 9 7 5 3 1

www.sterlingpublishing.com

Cover design by Igor Satanovsky
Book design by Sharon Jacobs

Your brain is remarkable.

Command central, mission control, the heart of the whole operation—for your entire life, your brain has managed mundane tasks and spectacular feats. Right this very moment, your gray matter is engaged, moving your eyes across the page, deciphering the words and sentences —all while managing to keep your heart beating and your lungs inhaling and exhaling.

But along with aching knees, age may claim some of your neurological sharpness. You lose your car keys more often or can't remember if you have dinner plans with your in-laws. Your brain is aging with the rest of you.

Scientists agree: there is no magic pill that will keep your brain working at peak condition as you age. A healthy diet and plenty of exercise are probably most important, along with staying socially active. While most neuroscientists don't agree that computer-based brain training is an effective way to keep the brain sharp, there is something to be said about learning something new or remembering a long-lost skill.

No matter your age, you need to stay cognitively active. And the more challenging the task, the more of a workout your brain gets. As a bonus, daily brain exercises and challenges can be fun. (That's why all major newspapers include a crossword puzzle.)

Your Daily Math was written to provide that daily cognitive workout. Each week, you'll have a full menu of mathematical problems, one per day, from seven categories.

IF YOU'RE RUSTY, DON'T WORRY.

The problems are easier at the beginning and get more difficult as the year progresses. And the questions aren't meant to stump you or make you feel inadequate. Even if it's been years since you solved for *x*, you can handle this math. The problems are designed to keep you thinking, putting together long-forgotten ideas with math that you do every day.

The point is not to pull up information that you learned decades ago. In these problems, the formulas are often provided. And each problem can be solved even if you have forgotten basic math rules.

The best part? There's no shame in getting a wrong answer. On each page, you'll find a (sometimes lengthy) explanation of the process for solving the problem. The goal is not to find the right answer. The goal is to think.

So what kind of math can you expect to do? Here are the seven categories:

NUMBER SENSE

Being nimble with numbers is what we math educators called numeracy. If you've got it, you can instinctively break a number down into various parts—including factors and sums. The foundation for numeracy happens in elementary school, but as grown-ups, we work on these skills throughout our lives, quickly estimating a sum or working with place value in decimals.

These problems focus on basic ideas about numbers and our number system—from the order of operations to recognizing factors and multiples. You'll also be encouraged to think about different kinds of numbers and how they work together.

ALGEBRA

Finding *x* has been a huge mathematical pursuit since the fathers of algebra, Diophantus and Al-Khwarizmi, began tinkering with equations in the third and ninth centuries, respectively. Contrary to popular belief, algebra was not invented to make your high school years a living hell, and you do use it in your everyday life.

In these problems, you're asked to solve for *x*, yes, but also to interpret lines on a coordinate plane and to simplify expressions and equations.

GEOMETRY

Unlike algebra, you've been doing geometry since you were teeny-tiny. (Remember Tinkertoys or Legos?)

Shapes dominate these problems, as you are asked to find the area, perimeter, and volume of figures. You will also explore transformations (translations, reflections, and rotations), as well as the geometry of the coordinate plane. (That's the *x-y* axis, in case you don't remember.)

APPLICATION

For most of us it's silly to do math just for math's sake. (Though it can be fun and interesting.) The real test of our deductive reasoning is using math in the real world.

And here's your chance. Using basic arithmetic, algebraic concepts, and geometry, you'll solve problems that you could encounter in your daily life. From personal finance to gardening to managing time, these challenges come from the real world.

PROBABILITY & STATISTICS

Being able to interpret statistical information and assess the likelihood of an event are perhaps the most useful math skills. Statistics are all around us, from our daily news to the reports some of us manage at work.

You'll apply critical-thinking skills to various scenarios involving probability and statistics. There are lots of questions about dice, cards, and raffle tickets. And you'll interpret graphs and statistical results.

LOGIC

For centuries, logic has been considered an important aspect of mathematics. Almost all of the mathematical concepts that we take for granted have been formally proven using logic. In this book, that's where things get really fun—and perhaps a bit more challenging.

These problems are not meant to be tricky or deceitful; however, they will challenge your ordinary way of thinking. Given a little bit of thought, you can find the answers. So don't give up too early.

GRAB BAG

Some problems just don't fit in any category. They often look like logic problems, but with a decidedly mathematical bent. A few come from math history— old problems and concepts that mathematicians have been playing with for centuries. And some are just downright silly.

In short, you won't know what's coming next with a grab bag question, but unlike logic questions, these problems will focus on numbers or shapes.

HOW SHOULD YOU APPROACH THIS BOOK?

There are 366 problems, one for every day of the year, plus leap year. The problems go in order by category. So if you work a problem a day, starting on a Monday, you'll solve a Number Sense problem every Monday.

When you don't remember how to do a problem, get creative. Draw a picture. Make a list of what you know. Create a table. Use a calculator. Look up a formula. There are no rules.

Just remember this: to get the most out of these cranial workouts, let your brain do the heavy lifting. Give yourself an opportunity to think through the problem before reaching for your smartphone or computer. Just think about how proud you'll feel when you find the answer.

So pick a day to start this journey. You can do a problem each morning, over lunch, or just before turning out the light. Over time, you'll remember more math than you thought you could. And you'll probably feel a little bit smarter because of it.

Week One

Our number system is in base 10. In other words, when counting, you reuse the numerals 0 through 9. Once you count to 10, a new place is created by adding a zero, such as going from 99 to 100. But this isn't the only base system that you use. Name the bases of the systems described below:

Months and years

Inches and feet

Seconds and minutes

Hours and days

Days and months

There are 12 months in a year, so the base is 12. Coincidentally, there are also 12 inches in a foot, so the base is 12 again. There are 60 seconds in a minute, so the base is 60. There are 24 hours in a day, so the base is 24.

But this last one is trickier. Some months have 30 days; some have 31. And February has either 28 or 29—depending on the year. So there is no consistent base for counting the days and months.

TUESDAY | **ALGEBRA**

Sandra has 15 chickens in her backyard coop. That's 4 more chickens than she had last week. Write an algebraic equation that describes Sandra's chicken situation. Remember, an algebraic equation requires 3 elements: a variable, an equal symbol, and some sort of operation (like addition or subtraction).

There are actually several different correct answers for this problem. It all depends on how you think about the problem itself. But first, identify what you don't know—that will be your variable. You don't know how many chickens Sandra started out with. You can choose whatever letter you'd like, but how about c for chicken? Now, what are you doing with that number? She's added 4 chickens to her coop, so one option is adding c to 4: $c + 4$. Now you know that the current number of chickens is 15, so you can simply add on an equal sign and 15: $c + 4 = 15$.

What are some of the other correct answers? First you can subtract c from 15 and leave 4 by itself on one side of the equation: $4 = 15 - c$. You can also simply rearrange the addition in the first solution: $4 + c = 15$. And of course, you can switch the sides of the equation: $15 = c + 4$ or $15 - c = 4$.

But not all rearranging works. For example, $c - 15 = 4$ is incorrect. That's because you are suggesting that Sandra subtract her current number of chickens from the number she had previously to find that she added 4 chickens.

WEDNESDAY | GEOMETRY

This triangle is equilateral. A line of symmetry is a line drawn through a figure that creates 2 mirror images. How many lines of symmetry does an equilateral triangle have?

Here's a cool way to think about lines of symmetry: Fold the figure, so that the 2 sides—one on each side of the fold—match up perfectly. In an equilateral triangle, there are 3 lines of symmetry, one through each of the 3 vertices. These bisect the opposite side, which simply means that they cut the opposite side into 2 equal parts.

Week One

Jackson has heard that he should not spend more than 5 times his annual salary on a house. His salary is $62,750, and he's considering purchasing a house that costs $337,750. Based on his rule, should Jackson purchase this house?

Jackson is facing a simple multiplication problem, but if he doesn't do the math, he could end up with a nasty financial surprise. All he needs to do is multiply $62,750 by 5 and compare the result with the price of the house: $62,750 × 5 = $313,750. This is well below the cost of the house. If Jackson wants to be fiscally responsible, he should keep looking.

Bertha's prize cantaloupes are growing like crazy. She's weighed all 6 of them and recorded the results (in pounds): 9, 10, 8, 12, 9, and 7. What is the mean weight of Bertha's cantaloupes (in pounds)?

Mean is the same thing as *average*. Just add all of the weights and then divide by the number of cantaloupes:

$$9 + 10 + 8 + 12 + 9 + 7 = 55$$
$$\text{and } 55 \div 6 = 9.2 \text{ pounds.}$$

Notice how the smaller cantaloupes bring the mean down. (Or stated another way, the larger cantaloupes bring the mean *up*.) A really small cantaloupe would skew the mean—and misrepresent the weights quite a bit.

For example, if the 7-pound cantaloupe were actually 3 pounds, the mean would be 8.5, which is quite a bit smaller than the larger cantaloupes in her collection.

SATURDAY | LOGIC

There are three Dalmatian puppies: Spot, Socks, and Patches. Spot has fewer spots than Socks, but more spots than Patches. Which puppy is the spottiest?

Since Spot has fewer spots than Socks but more than Patches, he has the middle number of spots. That means that Patches has the fewest number of spots, and Socks has the most spots.

Week One

How many lines can be drawn between these points? (Each line must be straight and cannot contain any bends, twists, or turns.)

Grab your pencil. Go ahead. If you don't feel that you can imagine this without drawing, mark up this book or sketch the circle and points on another piece of paper. Then start counting the lines. If you are careful, you'll find out that there are 10 lines that connect these 5 points.

It pays to be systematic, so that you don't miss a line or count one line twice. One way to approach the problem is by drawing all of the possible lines from each point. You'll quickly see a pattern: the first point creates 4 new lines, the second point creates 3 new lines, the third point creates 2, the fourth point creates 1, and the fifth point creates zero. Add them up and you get 10 lines.

MONDAY | **NUMBER SENSE**

What is the place value of each digit in 7,124?

There are several reasons that multiples of 10 are so wonderful. First off, they're easy to find—just slap one or more zeros on the end, and ta-da! You've got a multiple of 10.

But the other reason is that our number system is founded in these magical multiples. In other words, ordinary numbers are in base 10. This is the foundation of place value—or the general names of the locations of digits in a number.

Place value comes in a variety of flavors. To the right of a decimal point, each place value has *th* on the end—tenth, hundredth, etc. To the left of the decimal, the place value names lose the *th*: ones, tens, hundreds, etc. (Remember, when there is no decimal point, it's implied. In 7,124, the decimal is to the right of the 4.)

So the first step is to identify the decimal point. Then you can count to the left to find the place values with no *th* and to the right to find the place values with a *th*. Remember that there is a ones place to the left of the decimal, but there is no such thing as a *oneths* place. (Heck, that's not even a word.)

If you say the number out loud, the place value will become clear: seven thousand, one hundred twenty-four.

Here's another way to think of this: 7,000 + 100 + 20 + 4 = 7,124. So 7 is in the thousands place, 1 is in the hundreds place, 2 is the tens place, and 4 is in the ones place. And that's the story of place value.

TUESDAY | ALGEBRA

If $x = 8$,
what is $7 + x + 3$?

You're being asked to evaluate an expression. This boils down to sticking the 8 where the x is and then doing the arithmetic. Easy-peasy.

When you substitute, you'll get this: $7 + 8 + 3$.

Now just add to find the answer: $7 + 8 + 3 = 18$.

Week Two

WEDNESDAY | **GEOMETRY**

The figure below is an isosceles trapezoid. Two of the sides have the same measure.

How many lines of symmetry does this trapezoid have? (Remember, a line of symmetry divides a figure into 2 mirror images.)

If you think of the lines of symmetry as folds, you might try "folding" the trapezoid along opposite vertices or horizontally or vertically. That way, you'll find out quickly that an isosceles trapezoid has only 1 line of symmetry. Seems like there should be more, right?

THURSDAY | APPLICATION

You have run into the grocery store to pick up a few items for dinner. You have exactly $9 in your wallet and no credit card. (You cut that bad boy up a long time ago.) In your basket are a block of cheddar cheese ($3.59), a loaf of bread ($2.15), and a can of tomato soup ($0.95). Do you have enough cash?

Sure, you can add everything up and find the tax to get the exact total. But after a long day, who wants to do that? Because you only want to know if you have enough money, estimate. Round the price of each item to the nearest dollar and then add those whole numbers. Round $3.59 to $4; round $2.15 to $2; and round $0.95 to $1. Then add: 4 + 2 + 1 = 7. Your estimated total is $7. Even with tax, you'll have plenty of cash when you get to the register.

Qualitative data is descriptive; quantitative data is numerical; discrete data is quantitative data that can take on only certain values (like whole numbers); and continuous data can take on any value (like a range).

A dog shelter has many different kinds of dogs— mutts, purebreds, big dogs, little dogs, mature dogs, puppies. Give an example each kind of data that can be found at a dog shelter.

L ove these open-ended questions. Answers will vary, which allows for some creativity. But here's an example of possible answers. Qualitative data describes the dogs—spotted, shy, poodle, etc. Quantitative data counts or measures something—the total number of dogs, the total number of puppies, the lengths of the dogs' tails. Discrete data is restricted in some way, like whole numbers—the number of dogs, the ages of the dogs in years. Finally, continuous data is described in ranges—the dogs' weights or heights, the amount of dog food consumed each day, and so on.

So there are thousands of ways to describe the dogs in this shelter using data. Almost as many ways as there are different kinds of dogs.

SATURDAY | LOGIC

Four bullfrogs—Jerry, Elaine, Cosmo, and George—have taken up residence in a square house. Jerry is at the front of the house, while Elaine has settled near the back door. Cosmo is by himself on the east side of the house, and George is alone on the west side. Elaine switches places with George, who switches places with Jerry. Where is each bullfrog now?

When Elaine and George switch places, Elaine is on the west side and George is at the back. But then George switches places with Jerry, which puts Jerry at the back and George at the front. Cosmo didn't move anywhere, so he's still on the east side of the house.

SUNDAY | GRAB BAG

True or false? All numbers divisible by 6 are divisible by 3.
True or false? All numbers divisible by 3 are divisible by 6.

First question first. Start by rephrasing: If a number is divisible by 6, is it also divisible by 3? See if you can find a counterexample—an example that proves the statement is false. Twenty-four is divisible by 6 and also divisible by 3. Same is true for 12, 18 (of course), and even 162 (which is 6 × 27 and 3 × 54). It looks like there's a pattern here, but does it really hold up? Actually yes, and there's a mathy reason for it. Because 3 is a *factor* of 6, all numbers that are divisible by 6 are also divisible by 3. Another way to put it? Six is divisible by 3, so any number that is divisible by 6 is also divisible by 3.

But does this work the other way, too? If a number is divisible by 3, is it also divisible by 6? It's much easier to find this out by trial and error. All you need is one example that doesn't fit the statement. Twelve is divisible by 3 and divisible by 6. But 15 is divisible by 3 and is *not* divisible by 6. Insto-presto, you have your counterexample, and you know that the second statement is false.

Week Three

What is the place value of each digit in 3.257?

This number's decimal point is there in plain sight—and there are digits to the right of it.

Saying the number out loud may not be helpful in this case: Three and two-hundred fifty-seven thousandths. If you've forgotten about place value, you might have forgotten that you need to say "thousandths" at the end. Heck, you might have said, "Three point two five seven."

So you might have to think a little harder when there's a decimal point involved. Start with the 3, though. That's in the ones place—there are 3 ones in 3.257. The numbers to the right of the decimal point are *th* place values, starting with tenths. (Remember, *oneths* is not a word and not a place value.) So 2 is in the tenth place, 5 is in the hundredth place, and 7 is in the thousandth place.

By the way, what comes after the thousandth place? Why the ten-thousandth place, of course.

TUESDAY | ALGEBRA

If $x = 6$, what is $\dfrac{8x}{3}$?

You are evaluating the expression for $x = 6$. And because of the fraction, this expression looks difficult—when it's really not. Especially if you remember that the fraction can be described as division. So another way to write this problem is like this: $8x \div 3$.

Now all you need to do is substitute 6 for x, and then simplify: $8 \times 6 \div 3$. Now finish the arithmetic: $48 \div 3 = 16$. Not so difficult, right?

Week Three

WEDNESDAY | **GEOMETRY**

The perimeter of a square is 36 inches.
What is the area of the square?

For this problem, you need to really think about the characteristics of a square. In particular, a square has 4 sides with the same length. You also need to remember that the perimeter is the distance around the square. So if all of the sides are equal and the perimeter is 36, what is the length of each side? Divide the perimeter by the number of sides: $36 ÷ 4 = 9$ inches.

Now that you know the length of each side, you can find the area of the square. There are two ways to approach this. You can remember that a square is a rectangle, so you'll multiply the length by the width: $9 × 9 = 81$ square inches. Or you can remember that the area of a square is the square of one side: $9^2 = 81$ square inches. Either way, the area is 81 square inches.

THURSDAY | APPLICATION

As the treasurer of a brand-new nonprofit, you are in charge of the organization's finances. Over its first week, the nonprofit has received donations of $1,500, $3,000, $250, and $600. You also wrote a check for office supplies for $437, and a check for the office internet connection for $325. How much money did the nonprofit gain in its first week?

EXTRA CHALLENGE:

Find the answer *without* using a calculator.

Like most math problems, you can approach this question in different ways. For example, you can add all of the donations, and then add the checks. Finally, you can subtract the total of the checks from the total of the donations:
(1,500 + 3,000 + 250 + 600) – (437 + 325) = 5,350 – 762 = 4,588.

So the nonprofit has a little more than $4,500.

Here's another approach: Add the amounts of the checks and then subtract that total from the total of the first donation. Finally, add all of the rest of the donations. Or you can subtract one check from the first donation and the other check from the second donation. Then add the remaining donations.

Why do multiple methods work? Subtraction and addition can be done in any order.

FRIDAY | **PROBABILITY & STATISTICS**

Florence has surveyed her book club to find out what kinds of books people like to read. The results are in the table below.

NON-FICTION	ROMANCE	CLASSICS	SCIENCE FICTION	COMEDY
2	1	7	3	5

She's thinking of making a bar graph to display the information. What values would go on the vertical axis, and what values would go on the horizontal axis?

Okay, so this is kind of a trick question. Sorry. A bar graph can have either vertical bars or horizontal bars. So the answer depends on the type of bar graph that Florence creates. But one thing is absolutely true, no matter which way the bars are oriented: the number of books in each category determines the length of each bar.

Say Florence chooses a horizontal bar graph. In this case, the horizontal axis will show the frequency—or the number of books in each category—and the vertical axis will show the categories of books. The opposite is true for a vertical bar graph. In that case, the vertical axis shows the frequency, and the horizontal axis shows the categories.

Week Three

SATURDAY | LOGIC

At the theater, Elizabeth Taylor is sitting in her regular spot, seat D15. Marilyn Monroe is seated to her right in seat D16. Jayne Mansfield sits 3 seats to the left of Elizabeth Taylor, and Dorothy Dandridge is sitting next to Jayne Mansfield on her right. What seat number does Dorothy Dandridge have?

You know that Elizabeth Taylor and Marilyn Monroe are in seats D15 and D16. Since Marilyn Monroe is seated to Elizabeth Taylor's right, the seat numbers go up from left to right. This means that Jayne Mansfield is in seat D12 (15 – 3 = 12). So Dorothy Dandridge is in seat D13, because she is next to Jayne Mansfield on her right.

The real question is who is sitting between Dorothy Dandridge and Elizabeth Taylor? What a spot!

Week Three

Why are manholes round instead of square?

The answer to this requires a little bit of geometry. What is it about circles that makes them better than squares as manhole covers? The point of a manhole cover is to cover the hole, so that nothing falls in. But what keeps the cover from falling in? Turns out, it's the shape. If you had a square manhole cover you could tilt it along its diagonal—the line that connects opposite corners of the square—and the cover would fall in the hole. But a circle doesn't have these corners. No matter how you position a circular cover, it will never fall in the hole.

MONDAY | **NUMBER SENSE**

Round each number to the nearest tens place.

327.025

1,601.04

737.428

-248.96

Let's get one thing out of the way. The negative sign in the last number means nothing. Nada. It was just thrown in there to confuse you. Did it work?

Next, remember that rounding means chopping off the number at a particular place. But you can't do that chopping willy-nilly. The digit to the right of that place makes a difference in how you round. If that digit is 5 or larger, you'll round up. If it's less than 5, you'll round down.

The tens place is two digits to the left of the decimal. So in the first number above, the tens place is 2; in the second, it's 0; in the third, it's 3; and in the fourth, it's 4.

Considering the digit to the right of each of these numbers, you'll round this way: 330, 1,600, 740, and -250.

Remember, you don't need to round from right to left, from the last digit in the number to the tens place. Just consider the digit immediately to the right of the tens digit.

TUESDAY | ALGEBRA

Solve for x:
$$6 + x = 10$$

Ah, the mysterious x. Algebra may seem complicated, but there is actually a really simple way to solve for x in this problem. Ask yourself this question: What number when added to 6 gives you 10? More than likely the answer jumped out at you: $x = 4$.

Of course, there is a more formal way to solve for x. And this process comes in handy when solving more complicated algebraic equations. You want to undo the operation(s) in the equation. In this case, subtract 6 from both sides of the equation.

But why does this work? Subtraction is the inverse operation of addition. By subtracting 6 from the left side of the equation, you get x by itself. Because $6 + x = 10$ is an equation, you must subtract from both sides. If you only subtracted from one side, you would leave the equation unbalanced. When you subtract 6 from the left side, you get x. When you subtract 6 from the right side, you get 4. So $x = 4$.

Week Four

When a polygon is inscribed in a circle, each vertex of the polygon will lie on the circle, like this:

So when an equilateral triangle is inscribed in a circle, each of the three vertices is on the circle. What else is true about the vertices of this triangle?

There are actually a couple of things you can say about these vertices. It might be helpful to draw a picture. Make sure that the sides of the triangle are the same length. (An equilateral triangle has sides and angles of the same measures.) One thing to notice is that the vertices are the exact same distance from the center. That's because every single point on the circle is the same distance from the center. If you look at the arcs formed by the vertices—arcs are parts of the circle— you'll notice that these are the same length. So the vertices are equidistant from one another on the circle.

In fact, these facts are true for any regular polygon inscribed in a circle.

Week Four

THURSDAY | **APPLICATION**

Just after exercise, you check your pulse and find that in 10 seconds, your heart has beaten 22 times. How fast is your heart rate per minute?

First off, good for you for exercising. Now, the math. There are 60 seconds in a minute, but that's only part of the information that you need. How many 10-second intervals are in a minute? Just divide 60 by 10, and you have your answer: 6. So if your heart beats 22 times every 10 seconds, multiply by 6 to find out the number of times your heart beats each minute: 6 × 22 = 132. That's a pretty quick heart rate, which means you got a great workout.

Week Four

When you roll a fair, 6-sided die, what is the probability of getting a 2?

Probability means the likelihood that something will happen. Mathematically, this is the ratio of the favorable cases to all of the possible cases. (A ratio can be written as a fraction, with the first value as the numerator and the second value as the denominator.) In this situation, the favorable case is 1. That's because only one side of the die has 2 dots. The total possible rolls is 6, because the die has 6 sides. Therefore, the probability of rolling a 2 is 1 over 6 or $\frac{1}{6}$.

Week Four

There are five contestants in a pie-eating contest. Big Boy finished before Pie Man but behind Big Mama. Crusty finished before Ralph but behind Pie Man. In what order did the contestants finish?

One way to approach this problem is to arrange the names of the contestants as they appear in the problem. Take the second sentence first. Since Big Boy finished before Pie Man, but behind Big Mama, that order is Big Mama, Big Boy, and Pie Man. Next, look at the second sentence. Since Crusty finished before Ralph, that order is Crusty and Ralph. Finally, connect the two lists, using the last piece of information: Crusty finished after Pie Man. So, the second list appears *after* the first list. This means that the contestants finished in the following order: Big Mama, Big Boy, Pie Man, Crusty, and Ralph.

Week Four

SUNDAY | GRAB BAG

Myrtle had 10 children. Each of her children also had 10 children, who each had 10 children, who each had 10 children. How many great-great-grandchildren did Myrtle have?

A lot, that's for sure. To find the answer, it's easiest to find the number of children in each generation. The first generation is easy: 10 children.

Then her children had 10 children each: $10 \times 10 = 100$ grandchildren. Each of her 100 grandchildren had 10 children: $100 \times 10 = 1{,}000$ great-grandchildren.

Each of her great-grandchildren had 10 children: $1{,}000 \times 10 = 10{,}000$ great-great-grandchildren.

That's a big family.

Week Five

Round each number to the nearest hundredth place.

327.025

-1,250.677

130.0248

6,403.112

First, identify the hundredth place—which is two places to the right of the decimal point. So in the first number, you'll round the 2 to the right of the decimal point. Since the digit to the right of the 2 is 5, you'll round up to 3: 327.03. Rounding the next numbers to the nearest hundredth place gives you -1,250.68, 130.02, and 6,403.11.

Look at the third number more closely. Shouldn't you round the 4 to 5 and then the 2 to 3? This is a matter of convention, but most mathys will say to only consider the number directly to the right of 2.

Solve for *x*:
$$x - 8 = 24$$

This is another really simple algebra problem. Just ask yourself, "What minus 8 gives me 24?" That would be 32.

But of course there is a process for solving problems like these. To get the *x* by itself, you need to "undo" the subtraction by adding 8 to the left side of the equation. And whatever you do to one side, you must do to the other. So, you'll also add 8 to the right side. When you add 8 to the left side, you get *x* by itself. When you add 8 to the right side, you get 32. In other words, *x* = 32.

Week Five

WEDNESDAY | GEOMETRY

What is the measure of each angle of an equilateral triangle?

For this problem, you have to remember that the sum of the angles of a triangle is 180°. But what if you don't? There's a really clever way to find this out: "cut off" the angles of a triangle, and arrange these pieces so that the vertices of the triangle are at the same point. Here's a picture:

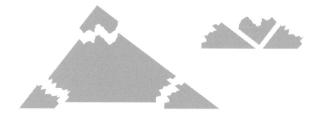

The three angles form a straight line, and a straight line has a measure of 180°. (How do you know this? Think of a circle, which measures 360°. You can divide a circle in half, using a straight line, and half of 360° is 180°.) This means that the sum of the measures of the angles of a triangle is 180°.

The sides of an equilateral triangle have the same length. The angles are also equal. So to find the measure of each of these angles, just divide 180° by 3. That means each angle is 60°.

Week Five

Keisha purchased a house for $213,000. Her monthly payments are $1,253, which she'll pay for 30 years. How much interest will Keisha end up paying?

Keisha may not want to know the answer to this question. Over the life of a mortgage, interest can add up fast. But here goes, anyway. Her monthly payments are $1,253, so Keisha will pay $1,253 × 12 or $15,036 per year. She has the mortgage for 30 years, so she'll pay $15,036 × 30 or $451,080 over the life of the loan.

Since the house originally cost $213,000, Keisha's interest payments will total $451,080 – $213,000 or $238,080. That's more than the original cost of the house.

FRIDAY | **PROBABILITY & STATISTICS**

What are the median, mode, and range of the data below? (The median is the middle number, the mode is the number that occurs most often, and the range is the difference between the largest and smallest numbers.)

16, 20, 16, 17, 16, 18, 17, 23, 16

The first step is to arrange this data in ascending (or descending) order: 16, 16, 16, 16, 17, 17, 18, 20, 23. Now it's really easy to pick out the middle number: the median is 17. The mode jumps out at you, too; 16 appears in the list four times. Find the range by subtracting the smallest number from the largest number: 23 – 16 = 7.

Notice that each of these represents the data differently. And think about this: What would happen if you substituted two of the 16s with 13s? Just goes to show that the data itself matters as much as the statistic that represents it.

SATURDAY | **LOGIC**

There are three people at the dinner table. Two are mothers, and two are daughters. How is this possible?

Thinking in mathematical terms, the data is not discrete. In other words, one of the women is both a mother and a daughter. So the women at the table are grandmother, mother, and daughter.

Week Five

SUNDAY | **GRAB BAG**

This morning, dog owners and their pooches have gathered at the local dog park. Some of the dogs have ticks. Eww! All in all, the people, dogs, and ticks have a total of 420 legs. There are twice as many dogs as people, and twice as many ticks as dogs. How many people, dogs, and ticks are there? (Assume that each person has 2 legs, each dog has 4 legs, and each tick has 8 legs.)

Just look at that little ecological system: the dogs depend on the people; the ticks depend on the dogs. Ain't nature grand?

But back to the problem. You have a few options—as usual. If you're big on writing equations, you can do that. There are 3 variables, so you'll need 3 equations.

On the other hand, if high school algebra is too darned far away, you can think of this in terms of ratios. What if there was only one person? That would mean there are 2 dogs, and 4 ticks, right? Counting up the number of legs, you're at 2 + 8 + 32—or 42 legs.

That number should look awfully familiar. There are a total of 420 legs, and with only 1 person, you have 42 legs. Just multiply 1, 2, and 4 by 10 to get the answer: 10 people, 20 dogs, and 40 ticks. For good measure, check the number of legs: 20 + 80 + 320 = 420 legs.

That's a lot of critters and a lot of legs.

Week Six

MONDAY | NUMBER SENSE

Put these numbers in ascending order:

327.23 | 327.32 | 326.32 | 327.232 | 327.231

Yep, those numbers look a lot alike—which is no accident. You'll need to look closely at the digits and place value to get the correct order. (You also need to remember that ascending order is from smallest to largest.)

So what's the smallest number in this list? Looking at each number from left to right will help you arrange the numbers. Start by looking at the hundreds place (or the first digit of each number). All of these numbers have 3 in the hundreds place, so that does not help. You're in the same spot with the tens place. All of these numbers have 2 in the tens place. But when you get to the ones place, things shake up a bit. The third number in the list has a 6 in the tens place. Since all of the other numbers have a 7 in the tens place, you've identified the smallest number: 326.32.

Now check out the numbers to the right of the decimal. In the tenth place of the remaining numbers, there's a 2, a 3, a 2, and a 2. Since 3 is larger than 2, you've identified the largest number in the list: 327.32. All that's left is to put the three middle numbers in order.

Since the remaining numbers all start with 327, just look at the numbers to the right of the decimal: 23, 232, 231. That makes this really easy, right? The remaining numbers in order are 327.23, 327.231, and 327.232.

TUESDAY | **ALGEBRA**

Solve for *x*:

$$6 = x - 4$$

By now, you may have gotten used to changing this algebraic problem into a simple question. But this time, the x is on the right side of the equation instead of the left. Does this matter? Not one bit. That's because of the commutative property for equality, which is a fancy-schmancy way of saying that $6 = x - 4$ is the same thing as $x - 4 = 6$. The question you're trying to ask is this:

What number minus 4 is 6? Clearly, the answer is 10.

And of course, you can approach this problem by performing the inverse operation on both sides of the equation. In this case, you'll add to both sides of the equation, since 4 is subtracted from x. Adding 4 to the left side of the equation gives 10. Adding 4 to the right side of the equation gives x. Therefore, $10 = x$, which thanks to the commutative property is the same thing as $x = 10$.

WEDNESDAY | **GEOMETRY**

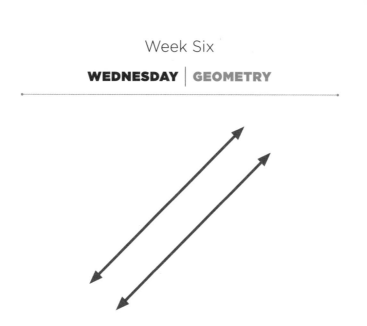

How can you tell for sure that these lines are parallel?

Parallel lines are lines that never intersect. (Technically, these lines must be on a plane—or a flat surface.) So one way to find out if these lines are parallel is to extend them forever and see if they intersect.

Not such a practical idea, right? Thankfully, there's another way. If the lines are parallel, they will have the same distance between them, no matter where on the lines you measure. But make sure that you're measuring the *perpendicular* distance. Between the two lines, draw a segment that forms a right angle with both lines. Then, repeat that process in another place between these lines. If these two line segments are the same length, then the lines are parallel.

THURSDAY | APPLICATION

Mr. Moneybags bought a piece of art for $4 million. He sold the piece for $7 million, and then years later bought it back again for $20 million. How much money did he make or lose on this piece of art?

Art sure can appreciate, and one thing is clear: Mr. Moneybags should have kept this piece in his collection. The easiest way to look at this problem is by thinking of each purchase as a loss of money (subtraction) and each sale as a gain (addition).

He purchased the piece originally for $4 million, which means he lost money. Then he sold it for $7 million—earning $3 million. But the last number is the kicker. Buying it back for $20 million is another loss, which means he lost a total of $17 million.

Hopefully, he can sell the piece again for more than that.

FRIDAY | PROBABILITY & STATISTICS

It has been a snowy winter in Bakersfield. The total snowfall in inches for each month is shown in the bar graph below. How much snow fell in all? How much more snow fell in February than January?

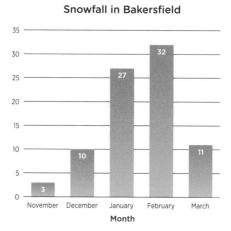

Snowfall in Bakersfield

To find the total snowfall, add the snowfall amounts for each month. These are indicated by the height of the bars. So, the total snowfall in November is 3 inches, December is 10 inches, January is 27 inches, February is 32 inches, and March is 11 inches. That's a whopping 83 inches for the year.

To find out how much more snow fell in February than January, subtract: 32 – 27 = 5 inches.

SATURDAY | LOGIC

A cat falls out of the window of a 20-story building and lives! How is this possible?

Right up front, you should know that the answer to this problem has nothing to do with cats landing on their feet. Nor does the problem assume that the kitty landed on something soft, like an awning or an overweight Saint Bernard. Nope, you need to pay attention to the phrase "20-story building."

If you automatically think that the cat fell out of the 20th story, then this is a magical cat. But notice that the question doesn't say this at all. The building is 20 stories tall, but the cat could have fallen out of a window on the ground floor—or if the cat is particularly good at landing gracefully on his feet, perhaps the second or third floor.

Week Six

Alejandro's age plus 3 is a perfect square. (A perfect square is the result of a number multiplied by itself.) His age minus 3 is the square root of that perfect square. (The square root "undoes" the perfect square. So, 81 is a perfect square, and 9 is its square root.) How old is Alejandro?

S tart with the first sentence. What number plus 3 gives a perfect square?

1 + 3 = 4; 6 + 3 = 9; 13 + 3 = 16; 22 + 3 = 25, and so on. From this list there are two options: 6 + 3 and 22 + 3.

Now think about the second sentence. What number minus 3 gives the square root of the first number?

1 − 3 = -2; 6 − 3 = 3; 13 − 3 = 10; 22 − 3 = 19; . . .

So Alejandro is 6 years old, because 6 + 3 is a perfect square and 6 − 3 is the square root of that perfect square.

MONDAY | NUMBER SENSE

Put the following numbers in descending order.

2,687,501.3

2,867,501.3

2,687,501.31

2,687,051.3

These numbers are huge. The commas and decimal points are going to help with the ordering. So is remembering that descending order is from largest to smallest.

In each number, there are seven digits to the left of the decimal point, so looking at each of these digits is useful. The first digit of each number—the millions place—is 2. Not helpful. But the second digits—the hundred-thousands place—are not all the same. Three numbers have 6 in the hundred-thousands place, and one number has 8 in that place. Clearly the one with the 8 is the largest number: 2,867,501.3.

Continuing from left to right in each remaining number, the next two digits are the same: 8 and 7. But things get interesting at the hundreds place. In two of the remaining numbers, that digit is 5, but in the third, that digit is 0. So you've found your smallest number: 2,687,051.3.

You're down to two remaining numbers: 2,687,501.3 and 2,687,501.31. These numbers are identical, except for one thing: that extra 1 in the second number, which makes it the larger of the two numbers.

TUESDAY | ALGEBRA

Solve for x.

$$\frac{x}{2} = 8$$

In this equation, x is being divided by 2 on the left side. So ask yourself: What number divided by 2 gives you 8? The answer is 16.

Of course, you can also choose to multiply each side of the equation by 2. This way, you get the x by itself. Why multiply? Because you're undoing the division on the left side of the equation. Multiplying the left side by 2 gives x. Multiplying the right side by 2 gives 16. So $x = 16$.

WEDNESDAY | GEOMETRY

A cross section is formed when a three-dimensional figure is cut by a plane. The cube below has been cut by a plane. (This means the plane and faces of the cube form right angles.) What is the shape of the two-dimensional figure created by this cross section?

The faces of a cube are squares. In addition, the adjoining sides of a cube are perpendicular to one another, and the opposite faces are parallel. Since the plane is perpendicular to the faces it cuts, it is also parallel to the faces it doesn't cut. Therefore, the cross section is a square—just like the faces of the cube.

THURSDAY | APPLICATION

Juniper's subscription to *Hey Girl!* magazine is about to expire. She has three options: **1)** subscribing for another year for $25, **2)** subscribing for two more years for $25, or **3)** buying her magazines at the newsstand for $4.50 each. How much will she save over the newsstand price if she gets the two-year subscription? How much will she save with a one-year subscription?

Clearly, the company wants Juniper to get a two-year subscription. And she'd be a fool not to do that. The savings are pretty substantial. If she purchased the magazine each month at the newsstand, she'd be paying $4.50 × 24 or $108. (Multiply the cover price by 24, or the number of months in two years.) Subtract $25 to find the savings, if she got a two-year subscription: $108 – $25 = $83. To find the savings with a one-year subscription, subtract the cost of two one-year subscriptions: $108 – $50 = $58.

That cover price has quite a markup.

Week Seven

The National Pie Association has taken a survey of people's favorite pies. They asked respondents to list their three favorite pies. The results are in the pie chart below. What is wrong with this graph?

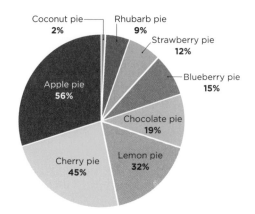

A pie chart is not the right graphic to use in this example. The biggest problem is that respondents were asked to list three of their favorite pies. For this reason, the percentages add up to more than 100%, a big no-no for pie charts. When respondents can offer answers in more than one category, a pie chart is not so delicious.

Week Seven

Ethan, Ava, and Madison made a dozen cookies each: one batch of chocolate chip, one batch of oatmeal, and one batch of peanut butter. Afterward, Ethan ate 1 chocolate chip and 2 peanut butter cookies. Ava ate 3 chocolate chip and 1 oatmeal cookie. Madison ate 1 oatmeal and 3 peanut butter cookies. How many of each kind of cookies are left?

A t the beginning of this little pig-out, there are 36 cookies in all—12 of each kind. Ethan and Ava have eaten a total of 4 chocolate chip cookies, so there are 8 of that kind left. Ava and Madison have eaten 2 oatmeal cookies, leaving 10 cookies. Finally, Ethan and Madison have eaten 5 peanut butter cookies, leaving 7 on the tray.

Use the smallest number of moves to change the triangle below so that it's pointed down instead of up.

This triangle actually represents a special kind of number called a *triangular number*. These are numbers that can be displayed by dots that create an equilateral triangle.

But that's not really the point. This triangle can be changed from pointing up to pointing down in only two moves. Take the bottom left dot and the bottom right dot and move them up to either side of the single dot at the top. Suddenly, the triangle is pointing in a completely different direction.

Week Eight

$$4 + 4 - 4 \times 0 = ?$$

Hello, Aunt Sally! If you don't get that reference, you might not have learned the common mnemonic for the order of operations: Please Excuse My Dear Aunt Sally. The order of operations tells you in what order to perform multiplication, division, addition, and subtraction. According to Aunt Sally, any operation inside parentheses is performed first (Please). Then come exponents (Excuse), multiplication (My), division (Dear), addition (Aunt), and finally subtraction (Sally). With the order of operations, you'd multiply 4 by 0, first. Then you can add and subtract:
$4 + 4 - 4 \times 0 = 4 + 4 - 0 = 8 + 0 = 8$.

Got it? Now let me throw a wrench in the situation. Multiplication and division can be done in any order. And addition and subtraction can be done in any order. So you don't always have to perform multiplication before division. And you are certainly welcome to subtract before you add. But if you're worried about getting things mixed up, you can simply follow Aunt Sally's lead to the letter.

Solve for x:
$$36 = 4x$$

This is another simple algebraic equation, so turning it into a question is a great option: 36 equals 4 times what? If you remember your times tables, the answer is pretty clear. Thirty-six equals 4 times 9.

And of course, you can approach this problem more systematically. To "undo" multiplication—and get the x by itself on the right side of the equation—just divide $4x$ by 4. As always, whatever you do on one side of the equation you must do on the other. So you'll also divide 36 by 4. This gives $9 = x$.

WEDNESDAY | GEOMETRY

An acute angle measures less than a right angle; an obtuse angle measures more than a right angle. Which of the following angles appear to be acute, and which appear to be obtuse?

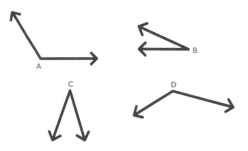

First, you need to remember that a right angle measures exactly 90°. This means that an acute angle measures less than 90°, and an obtuse angle measures more than 90°. You can certainly measure the angles above using a protractor. But you can also probably figure out which angles are acute and which are obtuse, just by thinking about what a right angle looks like. If one side of a right angle is absolutely horizontal, the other side will be absolutely vertical. So by that description, <A is obtuse and <B is acute.

The bottom two angles are a little trickier, if you're comparing to a right angle with one vertical side and one horizontal side. If you need to, turn the book so that one of the sides of <C is horizontal. You can see then that this angle is smaller than a right angle, so it's acute. Do the same with <D, and you can see that it is obtuse.

Week Eight

Your favorite football team is having an up and down game. It starts at the opponent's 25-yard line (25 yards from making a touchdown). In the next play, the team gains 6 yards. Then it loses 7. At what yard line is the team on now?

You can start by thinking about the plays individually. If the team is on the 25-yard line and gains 6 yards in the first play, it ends up on the 19-yard line. But how did that happen? Shouldn't you add? Not this time. Since the team is on its *opponent's* 25-yard line, it is moving closer to a touchdown or the 0-yard line. So when the team gains 6 yards, you'll subtract. But in the next play, the team loses 7 yards, so you need to add 7. This means that the team is at the 26-yard line or 1 yard upfield from where it started.

Another way to think of this is by finding the total yards won or lost in the two plays. Since the team gained 6 yards and lost 7, it lost a total of 1 yard. Losing a yard means adding, so the team is at 25 + 1 or the 26-yard line.

FRIDAY | PROBABILITY & STATISTICS

You are designing a graph that displays the annual donations to a charity over the last 15 years. What is the best graph—circle, bar, or line?

L ook closely at the data being measured. It doesn't really matter what the numbers are. Based on the type of data, you can make a smart decision.

You are displaying the total annual donations over 15 years. The likely message will be whether these donations have increased or decreased. A circle graph is typically used to show the percentage of the whole. You could make a circle graph that would show the percentage of the total donations given in each year.

A bar graph is used to compare. In this case, there would be 15 bars, one for each year.

Finally, a line graph allows data to be compared over time. The dots represent the annual donations for a given year. These are connected to show increases and decreases.

With all of that in mind, the line graph is the best choice. With it, you can easily see how donations changed from year to year, as the eye follows the line up and down.

Week Eight

SATURDAY | LOGIC

Five kids are playing reverse tag. In this game the person who is *it* tries not to be tagged by everyone else. Charlie is it. John is directly behind Charlie. Howard and Troy are side by side behind John. Oliver is behind Howard and Troy. John reaches for Charlie, but misses and falls. Troy trips and falls. Who tags Charlie first?

There are two people still standing: Howard and Oliver. Howard is between Charlie and Oliver, so Howard tags Charlie first.

SUNDAY | GRAB BAG

Your family tree has a branch for yourself, each of your parents, your grandparents, your great-grandparents, and your great-great-grandparents. How many branches are in your family tree? (Assume that each person has exactly 2 parents.)

If you're not careful, you might miss an important idea in this problem. It might be tempting to simply say that each generation has 2 people in it. But you have 2 parents and *4* grandparents and *8* great-grandparents and *16* great-great grandparents. In other words, for each generation, there are twice as many people as in the previous generation.

When you know the number of people in each of the generations, you can then simply add to find the total branches of the tree. Don't forget to add yourself.
1 + 2 + 4 + 8 + 16 = 31.

Of course, it might be easiest to simply draw the tree and then count the branches.

Week Nine

Put the following numbers in ascending order:

$$\pi \quad 3.27 \quad {}^5/_3 \quad 2.8$$

What's the smallest number in this list? Remember, π is the constant 3.14. . . . So both π and 3.27 have 3 in the ones place, so 2.8 is smaller. But is ${}^5/_3$ smaller than 2.8? Estimate by dividing 5 by 3. The ones digit is only 1, so you know that's your smallest number. That's followed by 2.8. And since π = 3.14 . . . , it's next, followed by 3.27.

TUESDAY | ALGEBRA

Solve for x:

$$3 + 2x = 11$$

What you have here is what algebra teachers call a two-step algebraic problem. Can you guess why? There are actually two operations: addition and multiplication. (Remember that when there is no space between a number and a variable, the number and variable are being multiplied.)

Some folks are able to find the answer just by thinking about the problem: 3 plus 2 times what number is 11? You can substitute various numbers for x to home in on the correct answer. For example, starting with $x = 3$, you get $3 + 2(3) = 3 + 6 = 9$. Since 9 is less than 11, you need to choose a larger number for x. How about 4? $3 + 2(4) = 3 + 8 = 11$. Bingo! So $x = 4$.

But if you don't want to go through all of that, just reach for your algebra rules. To get the x by itself, you want to undo two operations: addition and multiplication.

First get the $2x$ by itself, by subtracting 3. Then subtract 3 from the right side of the equation, which leaves $2x = 8$. Finally, divide each side of the equation by 2, which leaves $x = 4$.

Week Nine

A right triangle has one right angle. An obtuse triangle has one obtuse angle. And the sum of the measures of the angles of a triangle always equals 180°. Explain why a triangle can never be both right and obtuse.

Confused? Start by drawing a picture of a right triangle. Remember that a right angle measures 90° and an obtuse angle has a measure that is greater than 90°. That bit about the sum of the measures of the angles of a triangle is also a big deal. In a right triangle, the measures of the two non-right angles must add up to exactly 90°. (That's because 180° – 90° = 90°.) If these two angles add up to 90°, how can either of them be greater than 90°? Answer: they can't. And so you've shown that a triangle cannot be both right and obtuse.

Week Nine

THURSDAY | APPLICATION

A movie theater offers group rates on popcorn and drinks: $75 for popcorn and $50 for drinks for groups of 30. Movie tickets are $10 each (no discount). What is the total cost of popcorn, drinks, and tickets for a group of 30?

The cost of the popcorn and drinks is $75 + $50 or $125. Since each ticket costs $10 and there are 30 people, the total ticket price is $300. Add it all up, and you get a grand total of $425 for an evening at the silver screen with 29 of your best friends.

Week Nine

This chart shows the number of patients seen at a medical clinic, counted by age. This is not a bar graph. How can you tell? Why wouldn't you use a bar graph for this data? Which group is most represented? Which group is least represented?

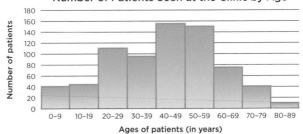

Number of Patients Seen at the Clinic by Age

This graph is called a *histogram*. Notice that the bars are touching each other; there is no space between them. And that's because the data is measured ranges. The youngest patients are between the ages of 0 and 9 years old. And the oldest patients are between the ages of 80 and 89 years old. A bar graph shows gaps between the categories, making it a poor choice for data measured this way.

Luckily a histogram can be read pretty much like a bar graph. The 40- to 49-year-old bar is taller than all the others, so the clinic sees more 40- to 49-year-old patients. But it sees fewer 80- to 89-year-olds than any other age range. (That bar is the shortest.)

Week Nine

SATURDAY | **LOGIC**

Gina has 5 sons. Each of her sons has 1 sister.
How many children does she have?

Did you think the answer was 10? If so, you overestimated. If each of her sons has 1 sister, there is only 1 sister in the family. So Gina has 6 children in all.

Week Nine

*The sum of two consecutive integers is 25.
What are the two numbers?*

Did you remember that an integer is a negative or positive whole number? The only way that two integers can be consecutive is if they are both negative or both positive. If you add two negative numbers, you will get a negative number. That means you're only looking for two *positive* consecutive whole numbers.

If you think of the numbers that add up to 25, you'll recognize that the consecutive numbers must be in the teens. One way to narrow this down is to think of the numbers that when doubled give a number close to 25: 12 + 12 is 24 and 13 + 13 is 26. In fact, 12 + 13 is 25. And you've found the correct answer.

Of course there are other ways to solve this problem. Since 25 has a 5 in the ones place, the two numbers must have values in the ones place that either add up to 5 or add up to 15. Look at those numbers that add up to 5 first: 0 and 5, 1 and 4, 2 and 3. Only 2 and 3 are consecutive, so the answer is 12 and 13.

Week Ten

$$(6 + 12) \div (4 + 5) = ?$$

Go from left to right, and you'll get one answer. Follow the order of operations, and you'll get the *right* answer. To claim victory, you've got to do the operations inside the parentheses first. Yes, even though that means adding before you divide.

So to solve this problem, start by adding 6 and 12 or by adding 4 and 5:
$(6 + 12) \div (4 + 5) = 18 \div 9$. Finally, divide to get 2.

What would happen if you ignored the parentheses and divided 12 by 4 first? Try it: $6 + 3 + 5 = 14$. A completely different answer.

And because of the order of operations, you know that it's the wrong answer.

TUESDAY | **ALGEBRA**

Write the following as an algebraic expression:

7 less than 8 times a number

You're asked to write an algebraic expression, not an equation. (No equal sign.) In this case, the math doesn't flow from left to right. "Less than" means subtraction, but you need to subtract the first value from the second value, not the second value from the first value. So start with the second value, which is "8 times a number." "Times" means multiplication and "a number" is any variable, so you can write this as $8x$. Then subtract 7: $8x - 7$.

WEDNESDAY | GEOMETRY

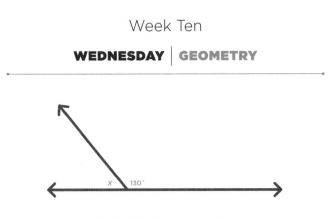

What is *x* in degrees?

Right away, you should notice that there are two *adjacent* angles. In other words, they share one side—the side that points up and to the left. Secondly, these angles form a straight line. And what's the measure of a straight line? That would be 180°. So *x* must be 180° – 130° or 50°. In fact, these angles have a special name. They're called *supplementary angles.*

THURSDAY | APPLICATION

Your sister sent you a birthday gift. It's beautifully wrapped with a ribbon, as shown below.

The box is 15 inches × 15 inches × 3 inches. There are 12 inches of ribbon in the bow. How many inches of ribbon did your sister use to wrap your gift?

Your sister sure does know how to dress up a present. To find the length of the ribbon she used, you need to consider how much ribbon was used along the length, width, and height of the box. Then add the 12 inches for the bow, and you have the answer.

The length and width of the box are the same, so you actually only need to find one of these and then multiply by 2. First, the ribbon is wrapped along the top of the box, down one side, then along the bottom of the box, and up the opposite side. Those lengths are 15 inches, 3 inches, 15 inches, and 3 inches. So the ribbon used one side of the box is 15 + 3 + 15 + 3 or 36. Since the dimensions of the width of the box are the same, she used the same amount of ribbon to wrap around the other side of the box. Now all that's left to do is add these, plus the length of ribbon used to make the bow: 36 + 36 + 12 = 84. So it takes 84 inches of ribbon to tie up that gift.

FRIDAY | PROBABILITY & STATISTICS

Ann is a movie reviewer. She's rated 6 movies on a 10-point scale: 8, 9, 8, 6, 7, and 4. But the publication she's writing for uses a 100-point scale. How can she change her ratings? How does this change affect the mean rating? How does it affect the median rating?

All Ann needs to do is multiply each of her current reviews by 10: 80, 90, 80, 60, 70, and 40. This is called the transformation of data. She's transformed the data into a different scale, by multiplying. And that's perfectly legal.

The mean (or average) of the original data is 7. (Remember, to find the mean, add all of the ratings and divide by 6, the number of movies.) When she changes the ratings to a 100-point scale, the mean is 70. In other words, multiplying by 10 also transforms the mean.

But what about the median? The median of the 10-point scale ratings is 7.5. (There is no middle number, so take the mean of the two middle numbers.) And the median of the 100-point scale ratings is 75. Again, the first median is multiplied by 10 to get the transformed median.

SATURDAY | LOGIC

Which of the below statements is true?

A. There is one false statement.
B. There are two false statements.
C. There are three false statements.
D. There are four false statements.

There are four statements in all. Since each statement says that there is a different number of false statements, only one statement can be true. And since only one is true, the other three must be false. That means the third option is the correct statement.

How can you connect all 9 of these dots, using only 4 straight lines and without picking up your pencil?

The trick to this puzzle is to recognize that you do not need to stay within the box created by the dots. With that in mind, here's one solution: Starting at the top left dot, draw a straight line down through the 3 dots in the first column. But instead of stopping at the last dot, continue the line below the bottom left dot. Then turn to create a diagonal line that goes through the bottom center dot and the right center dot, extending that diagonal so that it is outside the right column of dots and level with the top right dot. Then draw a line to the left that connects all of the dots on the top row. Finally, draw a diagonal that connects the top left dot, the center dot, and the bottom right dot.

There are other solutions, but all of them depend on extending the lines outside the box created by the dots. The moral? Think outside the box, of course.

-5 + 12 = ?

Don't remember the rules for adding integers? Don't worry. (But you do need to remember that integers are positive and negative whole numbers.) A number line can help you find the answer lickety-split. Draw a number line that extends from -10 to 10. (0 will be smack-dab in the middle.) Then locate -5 on that number line. Because you are adding a positive number, count 12 units to the right. The number you land on is 7, and that's your answer.

The rule is pretty darned handy. When adding numbers with opposite signs, first ignore the signs. Then find the difference (subtract). The answer will have the sign of the larger number. So in this case, you'll find the difference of 12 and 5, which is 7. The answer is +7 (not -7) because 12 is positive and larger than 5.

TUESDAY | ALGEBRA

Write an algebraic expression for the following:

The square of a number is added to 6 times the number and then divided by 2.

Translating words to numbers and mathematical symbols is the true art of mathematics. Otherwise, you wouldn't be able to apply math to the real world—which is kind of the point of having math around, anyway.

In this problem, you're asked to figure out which words in the sentence are numbers and which are symbols. You also need to find the order of the numbers and symbols. Because this is an expression, there is no equal sign.

Start with "a number." This is a clue that you need a variable. You're not told what number, so clearly the number is not known. You can choose whatever variable you want. The tried and true variable is x. First, you're asked to square that number: x^2. Next you need to add 6 times the number: $x^2 + 6x$. Finally, you are told to divide by 2. But do you divide $6x$ by 2, or x^2 by 2, or $x^2 + 6x$ by 2? The word "then" gives a good clue. After adding x^2 and $6x$, divide by 2: $(x^2 + 6x) \div 2$.

WEDNESDAY | GEOMETRY

What is the measure of the exterior angle of this triangle?

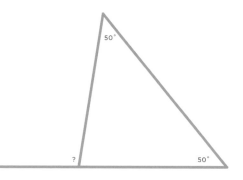

The exterior angle forms a straight line with one of the interior angles of the triangle. If you know the measure of the interior angle, you can just subtract from 180° to find the measure of the exterior angle. Problem is, you don't know what the measure of that interior angle is. Good thing it's pretty simple to find out.

The sum of the measures of the angles of a triangle is 180°. To find the missing angle in the triangle, add the two given measures and then subtract from 180°: 50° + 50° = 100° and 180° – 100° = 80°. Now you can subtract 80° from 180° to find the measure of the exterior angle: 180° – 80° = 100°.

You may have noticed something else, too. The measure of an exterior angle is the sum of the measures of the interior angles that are opposite the exterior angle. In this case, 100° = 50° + 50°. Yep, that works all the time.

THURSDAY | APPLICATION

On Tuesday, Melissa had 1,253 Facebook friends. She posted something controversial on Wednesday and lost 237 friends. Then she joined a group for her high school graduating class and gained 75 friends. (She was popular.) Over the next four days she lost 4 friends, gained 15 friends, gained 8 friends, and lost 1 friend. How many friends does she have now?

Oh, fickle Facebook. To find out how many friends Melissa has now, you can simply subtract the number of friends she lost and add the number of friends she gained. She started with 1,253 friends, and lost 237 friends: 1,253 – 237 = 1,016. After joining her high school graduating class group, she gained 75 friends: 1,016 + 75 = 1,091. Next up, she gains and loses friends by the day: 1,091 – 4 + 15 + 8 – 1 = 1,109. (Because she's adding and subtracting, she can simply go from left to right.)

And because math is so flexible, Melissa can approach this problem in another way: find the total friends she gained and the total friends she lost, and then add and subtract. She gained 75 + 15 + 8 or 98 friends. She lost a total of 237 + 4 + 1 or 242 friends. Now she can subtract and add again: 1,253 – 242 + 98 = 1,109.

This problem illustrates the associative property. You can group addition and subtraction in various ways and still get the correct answer.

Week Eleven

Aunt Crystal swears that her dog is going to win Prettiest Pup. She's sure, because on a local television station's website, a poll shows her Gertie winning by a landslide. But the actual voting takes place at the county fair. The day after the fair, Crystal is crushed to find out that Gertie lost—big time. How did the poll get the voting results so wrong?

There's not a lot to go on here, but since the results of the voting didn't match up with the results of the poll, you can assume one thing: there was a huge problem with the sample for the poll. A sample is a subset of the whole population (which in this case is anyone who voted at the county fair). The results of a survey or poll—or any research, for that matter—can only be generalized to the larger population if the sample is selected randomly.

An online poll is open to anyone who goes to the website. This excludes people who do not have access to the internet or who simply don't visit that particular website.

And it gets worse: an online poll may allow individuals to vote more than once, skewing the results even further. In addition, sometimes supporters pack the (virtual) ballot box with their votes. Finally, poll respondents may not even be planning to attend the county fair, much less vote.

No matter what you are polling, a large, random sample is required. And that's how Aunt Crystal's hopes were dashed.

All of my brothers except 2 are teachers.
All of my brothers except 2 are bakers.
All of my brothers except 2 are doctors.
How many brothers do I have?

There are two answers to this problem. In the first scenario, I have 3 brothers. One is a teacher, one is a baker, and one is a doctor. In the second scenario, I have 2 brothers. None of them is a teacher, baker, or doctor.

SUNDAY | **GRAB BAG**

A soda and a bag of chips cost $2. If the soda is $1 more than the bag of chips, how much does each cost?

I t's tempting to simply say that each costs $1, but that's not quite right. If the soda costs $1 more than the bag of chips, the bag of chips is free—and the total cost is not $2. What you're looking for are two numbers that add up to 2 but subtract to give you 1. And in this case, your answers are not going to be whole numbers. Unless you want to write two equations with two unknowns (yep, that works), you can use trial and error. Start with $1.25 and $0.75. These add up to $2, but when you subtract, you get 50¢, not a dollar. Now, the answer might seem obvious: $1.50 for the soda and 50¢ for the chips. Add them up, and you get $2. Subtract, and you get $1.

MONDAY | **NUMBER SENSE**

$$-13 + 4 = ?$$

Can't remember the rule? Picture or draw a number line. (Make sure it's long enough to include both numbers in the problem.) Find -13 on the number line, and then count 4 units to the right. (Why the right? Because you're *adding* a *positive* 4.) You'll end up at -9, which is the answer.

If you remember the rule, you can solve this problem without the number line. One number is negative and the other is positive, so ignore the signs, find the difference, and give the answer the sign of the larger number: 13 – 4 = 9, and since 13 is negative, the answer is -9.

TUESDAY | ALGEBRA

Three less than 5 times a number is 42.

Write and solve an equation.

For this problem, you need to translate the words into numbers, letters, and symbols. There's an "is," so you know you have an equation. Because "less than" is involved, it can be problematic to go from left to right.

For the heck of it, start with the variable. "Five times a number" is the same as $5x$. "Three less than" means to subtract 3 from $5x$: $5x - 3$. Now finish up with the equal sign and 42: $5x - 3 = 42$.

Finally, you need to solve for x or get x all by itself. Start by adding 3 to each side of the equation. That undoes the subtraction, giving you $5x = 45$. Now divide each side by 5, which leaves you with $x = 9$.

Week Twelve

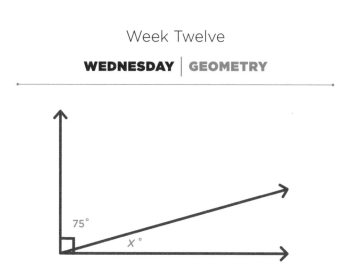

What is *x* in degrees?

This problem is similar to the supplementary angles problem from earlier. The angles are *adjacent*, but this time, they form a right angle. How do you know that? The little square in the corner of the largest angle. Because the largest angle is a right angle, you know that the smaller angles add up to 90°. So to find *x*, just subtract 75° from 90°: 90 – 75 = 15. So *x* = 15°.

Of course, you could also pull out your handy-dandy protractor and measure the angles, but honestly, do you even have one of those things anymore?

THURSDAY | APPLICATION

You have 5 U.S. coins that total 55¢. What are these coins?

The best approach to this problem is probably trial and error. Start with the coin(s) with the largest value. If you had a 50¢ coin, you'd need either a nickel or 5 pennies to make up 55¢. But that would give you either 2 coins or 6 coins, and you have 5 coins.

What about quarters? If you had 2 quarters, you'd need either a nickel or 5 pennies to make 55¢. Again, too few or too many coins. But 1 quarter might work. To get to 55¢, you could also have 3 dimes, but that's only 4 coins. How about 1 quarter, 2 dimes, and 2 nickels? That would give you 55¢ in just 5 coins.

FRIDAY | **PROBABILITY & STATISTICS**

A fair coin is tossed 7 times and comes up heads one of those times. (When a coin is *fair*, it's not weighted, so that one side will not come up more often than the other.) The coin is tossed again. What is the probability that it will come up heads this time?

This may be a very easy question or a very involved one, depending on how much you are concentrating. Flipping a coin is an independent event, which means that each time the coin is flipped, it has the same probability of coming up heads or tails. No matter how many times you flip a coin, the next time you will start over—meaning that the probability will always be $\frac{1}{2}$.

In the puzzle below, each row, each column, and each of the two main diagonals contains the letters found in the word SMART. What are the missing letters?

S	M	A	R	T
R		S		
	A	R		
		M		R
		T	S	

None of the rows can have more than one *S, M, A, R,* or *T*. The same goes for each of the columns and each of the diagonals. If you look carefully at the fourth row from the top, you'll notice that it has no *S*. But there are already *S*'s in each of the columns of that row, except for the second column. That means an *S* goes in row 4, column 2. Now all of the rows, except for row 3, have *S*s. And all of the columns, except for column 5, have *S*'s. Therefore, an *S* goes in row 3, column 5. Repeat this process for each of the different letters, and you'll find that the missing letters, in order by row and column, are: row 2: *T, M, A*; row 3: *M, T, S*; row 4: *T, S, A*; and row 5: *A, R, M*.

Week Twelve

SUNDAY | GRAB BAG

You have 15 coins. At least one is a penny, one is a nickel, and one is a dime. (There are no half-dollar coins.) The total value of the coins is 89¢. What are the coins that you have?

One way to solve this problem is by playing around with a pile of pennies, nickels, and dimes. Or you can take a look at the two numbers, 15 and 89, and see what you notice.

Because there's a 9 in the ones place of 89, you can deduce something about the pennies. There will be 4, 9, or 14 of Lincoln's coins. There cannot be 14, because then you have only 15 coins in all. And 9 doesn't seem likely—though it might be the right number. Start by considering 4 pennies.

If you have exactly 4 pennies, then there are 11 nickels and dimes (15 – 4 = 11). These nickels and dimes total 85¢ (89¢ – 4¢). What if there are 10 dimes? That would make 100¢ in dimes, which is more than 89¢, so that's a no-go. Try 5 dimes. This gives 50¢, which leaves 35¢ in nickels or 7 nickels. That's the right value of the coins (4¢ + 35¢ + 50¢ = 89¢), but there are too many coins (7 + 5 + 4 = 16).

You're so close, so how about trying 6 dimes? That's 60¢, which leaves 25¢ or 5 nickels. The value of the coins is 60¢ + 25¢ + 4¢ = 89¢. The total number of coins is 6 + 5 + 4 = 15. And so the correct combination of coins is 6 dimes, 5 nickels, and 4 pennies.

Week Thirteen

MONDAY | NUMBER SENSE

-6 + -9 = ?

There is a rule for adding two negative numbers, but you don't need it. Just picture (or draw) a number line. (Make sure this one is long, say -20 to 20.) Locate -6, and then count 9 places to the left. You're counting to the left because you're *adding* a *negative* number. You'll end up at -15, which is your answer.

But what's the rule? When adding two negative numbers, first ignore the signs. Then add the numbers and then make the answer negative. In this case, 6 + 9 = 15, but since 6 and 9 are both negative, the answer is -15.

Week Thirteen

TUESDAY | ALGEBRA

Twice a number increased by 4 is 24.
Write and solve the algebraic equation.

You can set up this problem from left to right. "Twice a number" means 2 times a variable: $2x$. "Increased by 4" means that you'll add 4: $2x + 4$. And then there's the little "is" and 24: $2x + 4 = 24$. And that's the equation.

Now for the solving part. Remember, to solve an algebraic equation you want to isolate the variable. Start by getting that 4 out of there. In other words, subtract 4 from both sides of the equation: $2x = 20$. Then divide each side of the equation by 2 to get rid of the 2 multiplied by x. This means that x is 10.

It's a good idea to check your work, which is pretty darned simple. Just substitute the answer into the equation. If both sides stay equal, you're good to go.

Week Thirteen

A stained glass window is a semicircle, as shown below. If each of the 3 glass pieces is identical, what is the measure of each of the three angles?

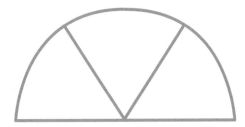

Because the whole piece is a semicircle, the angles add up to 180°. Since each section of the window is identical, you can divide 180° by 3. Therefore, each piece has an angle of 60°.

Bonus: What could you make with those three angles? A triangle! That's because a triangle has three angles with a sum of 180˚.

Week Thirteen

THURSDAY | APPLICATION

The corner grocery store sells three brands of liquid laundry detergent. SupaClean comes in a 64-ounce container and sells for $17.99. Sudz comes in a 150-ounce container and sells for $23.99. Whoosh! comes in an 80-ounce container and sells for $19.99. Which is the better deal?

I f you only have $18 in your pocket, you're obviously going to pick up the SupaClean brand. But if you want to get down and dirty with savings, it's smart to find the price per ounce.

Per means "divide," so this process is pretty simple. Just divide the price by the number of ounces. The lower the price per ounce, the better the deal. Start with SupaClean: $17.99 ÷ 64 ounces = $0.28 per ounce. Now, Sudz: $23.99 ÷ 150 = $0.16 per ounce. And finally, Whoosh!: $19.99 ÷ 80 = $0.25 per ounce.

The best deal is the biggest container, Sudz, at 16¢ per ounce.

Week Thirteen

The following stem-and-leaf plot shows the traffic fines for a city. How many infractions have fines between $40 and $70?

STEM	LEAF
2	5
3	7
4	5 8
5	0 5
6	0 5 8 8
7	5 5 8
8	0 0
9	0 5
10	0
11	5
12	
13	5
14	
15	0

A stem-and-leaf plot groups data by the leaf (the ones place) and the stem (the tens and larger places). With this kind of a plot, you can easily identify data that is grouped around a particular range in the tens place. In this case, you want to find the data that falls between $40 and $70. This means all of the data in the stem rows 4, 5, and 6. There are 8 numbers in those rows, so there are 8 fines that fall between $40 and $70.

SATURDAY | LOGIC

Olivia made some cookies. She ate 1 cookie and gave half of the rest to her sister. Then she ate another cookie and gave half of the rest to her brother. Olivia now has 5 cookies. How many did she start with?

The easiest way to solve this problem is by going backward. Olivia ended up with 5 cookies. Since she split her cookies with her brother, she must have had 10 cookies before. She ate 1 cookie, which means she had 11 cookies. And she split her cookies with her sister, so she started with 22 cookies. But remember, she ate 1 cookie at the very beginning. So Olivia started out with 23 cookies.

Week Thirteen

How many squares are in this drawing?

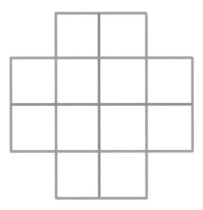

There are 12 small squares. That part is easy. But there are also some large squares that overlap. Start with the one at the bottom, which includes 4 small squares. Go up one row to find another 2 × 2 square. There's one more above and overlapping that one. Then there are three 2 × 2 squares arranged horizontally. That makes 6 squares made up of 4 smaller squares. Are there any squares made up of 9 smaller squares? (Those would be 3 × 3 squares.) Nope. So there are 18 squares in all.

MONDAY | **NUMBER SENSE**

$$3 - 12 = ?$$

In this case, you're subtracting two positive numbers, but the larger number is subtracted from the smaller one. Once again, a number line can come in handy. Locate 3 on a number line and count to the left (because you are subtracting) 12 spaces. The number you land on is -9, and that's the answer.

The rule for this problem is a little tougher. You need to know that subtracting a positive number is the same as adding a negative number. In other words, you can rewrite the problem as 3 + -12. Now you are adding a positive and a negative number, so you can ignore the signs, find the difference, and take the sign of the larger number. In short: 3 – 12 = 3 + -12 = -9.

Week Fourteen

TUESDAY | ALGEBRA

Twice a number plus 4 is equal to 12. Write an algebraic equation that describes this problem; then solve for x.

To write this equation, decode the math words. And for this problem, it's pretty simple to go from left to right—just like reading.

Start with "twice a number." "Twice" means "two times," so you'll be multiplying 2 by "a number." But what the heck is "a number"? That's your variable, which you can make whatever letter or symbol you'd like. So the first part of the equation is $2x$.

Now, you've got the phrase "plus 4," which clearly means to add 4. So add 4 to $2x$: $2x + 4$. Finally, you can translate the last phrase "is equal to 12." This is pretty darned straightforward: = 12. And the equation is $2x + 4 = 12$. When you solve this equation, $x = 4$.

The triangle below is an isosceles right triangle.
What are the measures of ∠1 and ∠2?

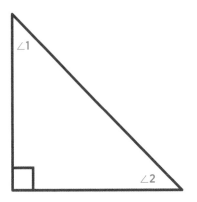

For this triangle to be isosceles, the two legs must have the same measure. (The legs of a right triangle are the two shorter sides that make up the right angle.) So that means that the angles that are not right angles have the same measure.

Because the sum of the measures of the angles of a triangle is 180°, the sum of these two non-right angles is 90°: 180 – 90 = 90. Divide by 2 to find out that the missing angles measure 45° each.

In fact, this is a really special triangle—called a 45-45-90 triangle—which plays a starring role in trigonometry.

Week Fourteen

THURSDAY | APPLICATION

It's 2:15 p.m., and you have a meeting with your boss at 4:00 p.m. There's so much still to do. How much time do you have to prepare for this meeting?

Finding *elapsed time* can be hard. And that's because one minute is 60 seconds and one hour is 60 minutes. In other words, time is in base 60, which is weird because our number system is in base 10. Talk about confusing.

Clearly what you want to do is subtract 2:15 from 4:00. (Since these times are both in the afternoon, you don't need to worry about the differences between a.m. and p.m.) But remember, time is in base 60. When you subtract 15 from 00, you're really subtracting 15 from 60, which leaves 45 minutes. To find the hours, you need to subtract 2 hours from 3 hours. (Not 4 hours. You're "borrowing" 1 hour from the 4 hours to get the 60 minutes.) That leaves 1 hour. So you have 1 hour and 45 minutes before your meeting.

If this approach doesn't work for you, try picturing a clock face. Starting at 2:15 p.m., imagine the minute hand moving around the clock until it reaches 12. In that time, the hand has moved $3/4$ of the way around the clock, or 45 minutes. At the same time, the hour hand has moved to 3. Now imagine the minute and hour hands moving around the clock until they reach 4:00. One hour has passed, and the total elapsed time is 1 hour and 45 minutes.

Week Fourteen

FRIDAY | **PROBABILITY & STATISTICS**

By the fourth quarter of school, Tamera has an 88 average in English class. That's only two points from an *A*. What grade does she need to earn in the fourth quarter to get a 90 for the year? (Assume that each quarter is weighted evenly.)

If Tamera had all of her grades for each quarter, she would add them together and then divide by 4 (the number of quarters). She'll use that process—but with a variable—to find out what she needs to earn in the last quarter.

Since each quarter is weighted evenly, she can assign 88 as the average for each of the first 3 quarters. She doesn't know the last quarter, but she can call it x. The first step in finding the mean is to add up all of the values: $88 + 88 + 88 + x$. Then divide by the number of values: $(88 + 88 + 88 + x) \div 4$. In this case, there's another step. She wants her final grade to be 90, so she can turn this expression into an equation: $(88 + 88 + 88 + x) \div 4 = 90$.

Now she can simplify the left side of the equation and then solve for x. An easy first step is to add up the three 88s: $(264 + x) \div 4 = 90$. Since the addition is in parentheses, multiply each side of the equation by 4: $264 + x = 360$. Now subtract 264 from both sides of the equation: $x = 96$.

So Tamera has to do very well this quarter—earning a 96 average—in order to get an *A* for the year.

SATURDAY | LOGIC

Yesterday is 4 days before the day after Sunday.
What is tomorrow?

Take a look at this problem backward. What is the day after Sunday? Monday. And what is 4 days before Monday? That would be Thursday. So yesterday is Thursday, which means today is Friday and tomorrow is Saturday.

Week Fourteen

SUNDAY | GRAB BAG

Which of the following pairs of numbers does not belong?

53 and 36
59 and 43
39 and 22
34 and 17

You can approach this problem in a number of different ways. Consider what the numbers have or don't have in common, for example. In that way, you might notice that 59 and 43 is the only pair of prime numbers. (Prime numbers are divisible by only two numbers: 1 and the prime number itself. So 2, 3, 5, and 7 are prime numbers.) And you would be right. Or you may have noticed that 59 and 43 is the only pair of odd numbers. Right again.

And there's another reason that 59 and 43 is the oddball pair. If you subtract the second number from the first in each pair, you'll get 17 for the first, third, and fourth pairs—but not for the second pair. In that case, you get 16.

MONDAY | NUMBER SENSE

$$-3 - 8 = ?$$

Now these addition and subtraction problems are getting really complicated. Of course, using a number line is fail-proof. Locate -3 on the number line and then count 8 spaces to the left. Why the left? Because you're subtracting. You'll end up at -11, which is the correct answer.

Now for the rule: first, you need to remember that subtracting a positive number is the same as adding a negative number. So you can rewrite the problem as -3 + -8 = ? To add two negative numbers, ignore the signs and add: 3 + 8 = 11. Then make the answer negative. That means you'll get -11.

TUESDAY | ALGEBRA

Solve for x.

$$\frac{x}{3} = \frac{6}{9}$$

Perhaps you can look at this problem and see what x is. The two fractions must be equal, so $x = 2$. But what if you didn't notice that solution? There's an easy process you can follow.

The equation above is called a proportion—two ratios (written as fractions) that are equal. To solve a proportion, cross multiply and then solve for x. *Cross multiplying* means multiplying along an X that links the numerator of one fraction to the denominator of the other fraction, and vice versa. So you'll multiply x and 9 to get $9x$ and then multiply 3 and 6 to get 18. Set these two equal: $9x = 18$. To solve for x, just divide each side of the equation by 9, so $x = 2$. And there you have it: solving a proportion using cross multiplication.

Week Fifteen

What are the missing measures of the angles
in the figure below?

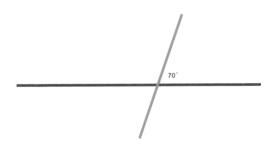

The angle measuring 70° forms a straight line with the adjacent angle to the left. This means that the sum of those two angles is 180°. So the measure of the angle to the left of the 70° angle is 180° – 70° or 110°. The 110° angle and the one below it form a straight line, so the sum of those two angles is 180°. Therefore, the measure of the angle below the 110° angle is 70°. Lastly, the original 70° angle and the one below it also form a straight line. Therefore, the measure of the angle below the original 70° angle is 110°.

Turns out there are only two different measurements for all four angles.

Week Fifteen

THURSDAY | APPLICATION

A super-couponer is shopping on double-coupon day. Each of his coupons is worth twice as much as the face value. Woo-hoo! He has the following coupons: twelve 50¢-off coupons, seven 25¢-off coupons, five 10¢-off coupons, and eight $1.50-off coupons. How much will he save in all?

So what would you like to do? Double the coupon value first or find the savings and then double? Either way works just fine.

How about finding the total savings of all of the coupons and then doubling the value? Start with the 50¢-off coupons. There are 12 of those, so the total savings (before doubling) is 600¢ or $6.00. Do the same with the remaining coupon values: 7 × 25¢ = 175¢ or $1.75; 5 × 10¢ = 50¢ or $0.50; and 8 × $1.50 = $12.00. Now add: $6.00 + $1.75 + $0.50 + $12.00 = $20.25.

But that's not the total savings. Remember, it's double-coupon day, so this smart shopper will save a whopping $40.50. That's quite a savings.

Week Fifteen

Last year, a neighborhood's median house sales were, by quarter:

Q1: $255,000
Q2: $270,000
Q3: $275,000
Q4: $260,000

This year, they are:

Q1: $265,000
Q2: $295,000
Q3: $297,000
Q4: $285,000

What is the best graph to describe this data: a simple bar graph, a two-bar graph, or a circle graph? Explain why.

In essence, you want to describe two changes: the change in home prices from quarter to quarter and the change from year to year. A circle graph is not going to be the best option, because it describes the relationship of parts to the whole.

So you really want some kind of bar graph. The simple bar graph will show 8 bars, arranged from oldest data (Q1 last year) to newest data (Q4 this year). This will allow you to compare data from quarter to quarter, but not from year to year. Instead, use a two-bar (or grouped bar) graph. In this type of graph, the data from Q1 last year and Q1 this year is put side-by-side. This allows you to compare the data year over year.

SATURDAY | LOGIC

Five magazines are stacked on a table. The *New Yorker* is placed below *Mad* magazine. *Esquire* is placed above *Playboy*. *National Geographic* is placed below the *New Yorker*. And *Playboy* is placed above *Mad*. Which magazine is at the bottom of the stack?

The *New Yorker* is below *Mad*, so *Mad* cannot be on the bottom of the stack. Likewise, *National Geographic* is below the *New Yorker*, so the *New Yorker* cannot be on the bottom of the stack. That leaves *Esquire*, *Playboy*, and *National Geographic*. *Playboy* is above *Mad*, so it cannot be at the bottom of the stack. And *Esquire* is above *Playboy*, so it cannot be at the bottom of the stack. That leaves *National Geographic* at the bottom of the stack.

In fact, the magazines are in this order, from bottom to top: *National Geographic*, the *New Yorker*, *Mad*, *Playboy*, and *Esquire*.

Week Fifteen

SUNDAY | GRAB BAG

What is so special about the number 8549176320?

The first thing you may have noticed is that none of the digits in this number repeats. There is only one 8, only one 5, and so on. Next, you may have noticed that each single digit number from 0 to 9 is represented. But why are the digits in this order? In this situation, the answer has nothing to do with math. The last digit is the biggest clue. If you think of the words for each of these numerals, you'll find the answer quickly. They're in alphabetical order. So simple, but so difficult to see.

Week Sixteen

MONDAY | NUMBER SENSE

$$4 - (-7) = ?$$

First off, notice those parentheses around -7. They are there so that the subtraction symbol doesn't get in the way of the negative sign. In other words, they only help clarify the notation.

You're subtracting a negative number from a positive number. Number lines to the rescue! But this time with a twist. Locate 4 on the number line and then count 7 . . . but in which direction? If you were subtracting a positive number, you'd count to the left. And if you were adding a negative number, you'd also count to the left. So if you're subtracting a negative number, you'll count to the *right*. (Grammar geeks will recognize this as a double negative, which is always positive.) You'll land on 11.

Rule time. When you subtract a negative number, it's the same as adding a positive. So you can rewrite the problem and solve: $4 - (-7) = 4 + 7 = 11$. Voilà!

TUESDAY | ALGEBRA

Solve for x.

$$\frac{12}{6} = \frac{4}{x}$$

To solve this proportion, use cross multiplication. *Cross multiplying* means multiplying the numerator of the first ratio by the denominator of the second ratio and then multiplying the denominator of the first ratio by the numerator of the second: $12x = 24$. Finally, divide each side to find that x is 2.

There is another way to solve this problem. (There usually is in math.) Notice that the first ratio can be turned into a whole number: $12 \div 6 = 2$. What value of x would give you a 2 on the right side of the equation? (Of course it's the same answer as above.)

Week Sixteen

WEDNESDAY | GEOMETRY

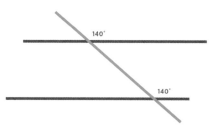

In the figure above, the two horizontal lines are parallel.
Find the measures of all of the angles in the figure.

If you remember your high school geometry, you can find the measures of these angles using a few theorems.

Or you can just recognize the angles that form straight lines and then find the measurements. For example, in the top four angles, the 140° angle forms two straight lines—with the angle on the left and the angle below it. That means that the smaller angles are 180° − 140° = 40°. But what about the other larger angle? That one is also 140°, because it forms straight lines with each of the two smaller angles.

The same is true for all of the angles in the bottom part of the figure—and for the same reasons. They form straight angles with one another, so you merely need to subtract from 180° to find their measures. So all of the larger angles in the figure measure 140° and all of the smaller angles measure 40°.

But what about those theorems? In this figure, there are vertical, corresponding, alternate interior, alternate exterior, and interior angles on the same side of the transversal. If you remember what these pairs of angles look like and when they are congruent, you can find all of the angle measures when given only one angle measure.

Week Sixteen

THURSDAY | APPLICATION

You had $637.95 in your bank account. Your mortgage payment was automatically withdrawn, leaving you with -$532.15. As a penalty for the overdraft, the bank charges a flat fee of $50 plus 3% of the overdrawn amount. How much do you need to deposit to bring your account into the black?

Did you notice? Not all of the information given is necessary. In fact, it doesn't matter what you had before the mortgage payment was withdrawn. What matters is how much you are overdrawn. To get back to positive, you need to deposit $532.15. That will bring your account to zero.

But the bank is also going to withdraw $50, plus 3% of the overdrawn amount. To find the percentage, multiply: 0.03 × 532.15 = $15.97.

Add the $50: $15.97 + $50 = $65.97. Add that to the $532.15 that you already deposited to get $598.12.

Ugh. No one wants to spend money on that.

FRIDAY | **PROBABILITY & STATISTICS**

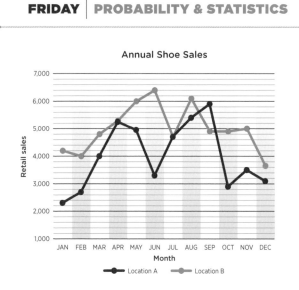

This graph shows the sales of shoes from two store locations. In what month did Location A have its highest sales? In what month did Location B have its highest sales? Which location has a better sales record overall?

Location A's sales are indicated by the blue line, and Location B's sales are indicated by the orange line. So Location A had its highest sales in September—back to school. Location B had its highest sales in June—at the start of summer vacation. Overall, Location B has a higher sales record, since the red line is above the blue line except in April, July, and September.

But if you want to be sure, add the monthly sales for Location A and compare with the sum of the monthly sales for Location B.

Week Sixteen

SATURDAY | LOGIC

Anna, Simone, Cora, and Carrie are on sports teams. Anna and Simone play lacrosse, and the others play soccer. Simone and Carrie are on the swim team. Everyone, except Anna, plays tennis. Which girl(s) is on the soccer and tennis teams?

The easiest way to approach this problem is by making a list of the sports that each girl plays. Anna plays lacrosse. Simone plays lacrosse and tennis and is on the swim team. Cora plays soccer and tennis, and Carrie is on the soccer, tennis, and swim teams. So this means that Cora and Carrie are both on the soccer and tennis teams.

SUNDAY | **GRAB BAG**

Complete the squares so that each of the rows, columns, and diagonals has a sum of 15. Each number, from 1 to 9, should be used only once. Use only positive numbers.

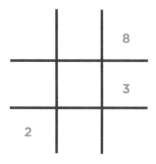

This is kind of like one-ninth of a Sudoku puzzle, except that the sums of the rows, columns, and diagonals must be 15. You have the following numbers to place: 1, 4, 5, 6, 7, and 9. There are two places that are easy to fill in. The center square must be 5, since 2 + 8 = 10. And the bottom right square must be 4, since 8 + 3 = 11. Now you can fill in the bottom center square with a 9 (2 + 9 + 4 = 15). And the left middle square is 7 (7 + 5 + 3 = 15).

In the top left, the missing number is 6 (6 + 7 + 2 = 15). And the top middle square must be 1—the only number not used. It's a good idea to check the top row (6 + 1 + 8 = 15) and the other diagonal (6 + 5 + 4 = 15) to make sure there are no errors.

MONDAY | NUMBER SENSE

-9 – (-4) = ?

The addition and subtraction is getting really complicated now. In this problem, you're subtracting a negative number from a negative number. Yikes! Again, a number line can be useful. Locate -9 and then count 4 to the right. Why the right? Remember that subtracting means counting to the left *and* negative numbers mean counting to the left. Therefore, you count to the right. You'll end up at -5.

What rule can you use to solve this problem? Subtracting a negative is the same thing as adding a positive.

Rewrite the problem and solve: -9 – (-4) = -9 + 4 = -5. (When adding a negative and a positive, ignore the signs, find the difference, and take the sign of the larger number.)

Week Seventeen

TUESDAY | ALGEBRA

If $x = -4$, what is $x^2 + 2x + 1$?

Ooh! Tricky! You probably know by now that this is a simple substitution problem, right? But you're substituting a negative number into an expression with one of those tiny 2s.

The tiny 2 is just an exponent—a square, to be precise. It means that you'll multiply x by itself: $x^2 = x \times x$. When you substitute, be sure to put parentheses around -4. This way it's clear that you're squaring and multiplying by -4 and not 4: $(-4)^2 + 2(-4) + 1$.

There's a lot going on there, but the order of operations will help you out. Exponents go first: $16 + 2(-4) + 1$. Next, multiply 2 and -4: $16 + -8 + 1$. Now add from left to right: $8 + 1 = 9$. And that's the value of this expression when $x = -4$.

WEDNESDAY | GEOMETRY

What kind of three-dimensional figure can be made from the two-dimensional shape below?

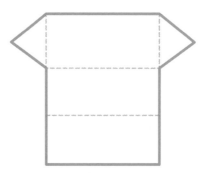

This kind of figure is called a net. It's what you get when you unfold a three-dimensional figure into a one-dimensional figure. If you've ever worked at a pizza shop, you know this idea well. Pizza boxes are shipped flat and folded at the shop to create the three-dimensional boxes. In this case, there are 3 rectangular faces and 2 triangular faces. This means that the net makes a triangular prism.

But if you didn't remember that term, you can just describe it: a three-dimensional figure with 2 opposite faces that are triangles, and 3 rectangular faces. Or you can remember that this is what a pup tent looks like.

Week Seventeen

THURSDAY | APPLICATION

A teacher has $300 to spend on a field trip to a museum. The bus fee is $150, and each admission ticket for the museum is $10. How many students can go?

If you're really math savvy, you can set up an equation and solve: $300 = 10x + 150$. But where does that equation come from? The teacher has $300 to spend, so 300 will be on one side of the equal sign. (Honestly, it doesn't matter which one.) The variable represents the number of students, and each must pay $10 to get into the museum. Then on top of the admission charge, the teacher needs to pay for the bus, which is $150. To solve this equation, subtract 150 from each side to get $150 = 10x$. Then divide each side by 10, which leaves $x = 15$ students.

Of course, you can do the arithmetic without even creating an equation. Subtract $150 from the $300 budget to see how much is left for the museum admission. The teacher has $150 to get her students into the museum. Since each ticket costs $10, 15 students can go on the trip.

Week Seventeen

A recent headline reads: "New research shows that Americans have a 25% chance of having a heart attack." What additional information do you need to understand the research?

A headline can only tell you so much. In this case, you have no idea who conducted the research or whether the work was done responsibly. But assume that the researchers knew what they were doing and were unbiased.

The first question you might ask is whether this statistic applies to *all* Americans. Do babies have a 25% chance of having a heart attack? What about teenagers? More than likely, this research focused on a particular age group. In addition, were the subjects active and eating a good diet? What about their stress levels?

The sample of the population studied has a big effect on how the research can be generalized. If researchers only looked at seniors who are moderately active and eat a regular diet of processed foods, the results cannot be generalized to 20-something health-nuts who are training for marathons.

Week Seventeen

SATURDAY | **LOGIC**

Crud. You left your calendar at home, and now you have no idea when your appointments are. Each one is an hour apart, starting at 10:00 a.m. You remember these details:

Your haircut is after your dental cleaning and coffee with your best friend. The coffee is before the dental cleaning. And you have to pick up your dog from the groomer after your haircut.

Starting at 10:00 a.m., when are each of your appointments?

It makes more sense to start out with the second bit of information—that your coffee date is before your dental cleaning. (And isn't that a good thing? Coffee after a dental cleaning sounds gross.) Then you can place the haircut after the dental cleaning. Finally, picking up the dog is last. (Another stroke of genius, since you can go straight home after that.) This means that the coffee is at 10:00 a.m., the dental cleaning is at 11:00 a.m., the haircut is at noon, and you'll pick up the pooch at 1:00 p.m. Now, to get everywhere on time.

Week Seventeen

What are the missing numbers that make the sum correct?

This is just addition, so start at the right: 5 + 6 = 11. That means 1 is the missing number on the bottom. But you'll have to carry a 1 to the next place value: 1 plus 8 plus a number with 6 in the ones place. The options for the sum are 6, 16, 26, 36, 46, 56, and so on. But only 16 makes sense: 1 + 8 + 7 = 16. So 7 is the missing number in the second row. You'll need to carry a 1 again: 1 plus what plus 2 gives 9? You know for sure that the sum is 9, since there are no other digits to add. Therefore, the missing number on the top must be 6: 1 + 6 + 2 = 9.

To summarize: 685 + 276 = 961. All the missing numbers have been found.

$$-10 - (-10) = ?$$

In this case, the answer may be pretty obvious. Negative 10 minus negative 10 equals what? When you subtract any number from itself, you get zero. Every time. So the answer is 0.

But you can also think of this in terms of the number line or a rule. On the number line, locate -10. Then count 10 places to the right. (Remember, when subtracting a negative, you count to the right, not the left.) You'll end up at 0.

Of course, when you subtract a negative number, it's the same thing as adding a positive number. Therefore, you can rewrite the problem and solve:
$-10 - (-10) = -10 + 10 = 0$.

If $x = 1$, what is $x^2 - 8x + 7$?

Now you've got a negative number to substitute into an expression with a square *and* subtraction. Start out by substituting, being sure to use parentheses: $(1)^2 - 8(1) + 7$. Now evaluate—which is a fancy way of saying *do the arithmetic*: $1 - 8 + 7 = -7 + 7 = 0$.

Whew! Did you get all of that? When you subtract 8 from 1, you get -7. Adding a -7 and +7 is 0.

What is the volume of the cylinder below? The height is 12 and the diameter is 4. (The formula for the volume of a cylinder is $\pi r^2 h$.)

There are three letters in the formula—π, r, and h. Pi is actually not a variable but a constant that represents 3.14. . . You know the height is 12, but you don't know the radius. Instead you've been given the diameter, which is twice the length of the radius. Therefore, the radius is 2.

Now you can plug everything into the formula and find the volume of the cylinder: $3.14 \times 2^2 \times 12$. First evaluate the exponent: $3.14 \times 4 \times 12$. Now multiply to get 150.72 cubic units. (Why cubic units? Because volume measures three dimensions.)

Week Eighteen

THURSDAY | APPLICATION

Four people share a house. Pablo has the largest bedroom with an adjoining bath. Vincent has the next largest bedroom with his own bath (but it's in the hall). Henri and Andy are downstairs with similar-sized rooms and share a bath. The roomies decide that based on the sizes of their bedrooms and the proximity of the bathrooms, Pablo should pay $3/7$ of the rent, Vincent should pay $2/7$, and Henri and Andy should each pay $1/7$. If the rent is $840, how much will each person pay?

It may seem easier to change the fractions to decimals, but actually, there's an even simpler process. Notice that everyone is paying a multiple of $1/7$. Find $1/7$ of the rent, and you'll have Henri's and Andy's rent. And $1/7 \times \$840$ is the same thing as $\$840 \div 7$, which equals $120.

To find $2/7$, double $120: 120 \times 2 = \$240. That's Vincent's rent. And finally, triple $120 to find $3/7$ or Pablo's rent: $120 \times 3 = \$360.

Of course, you can always multiply $840 by the numerator of the fraction, and then divide by 7 (the denominator). Once again, mathematics shows its flexibility.

FRIDAY | PROBABILITY & STATISTICS

What can you say about the graph below?

SAT Scores by School

Honestly? Not much at all. There is a huge problem here: the horizontal and vertical axes are not labeled. It can be assumed that each bar represents a school's SAT scores, but there's no indication that these are total scores, average scores, or even median scores. And because none of the bars are labeled, you can't know which school is which.

And there's more: because the vertical axis is not labeled, there's no way to know what each of the horizontal lines means. Is the height of each bar measured by ones, tens, or hundreds?

Without labels on the vertical and horizontal axes, a bar graph is worthless.

Week Eighteen

SATURDAY | LOGIC

Mrs. Brown, Mr. Black, and Miss White
are meeting for the first time.

"How amazing," the man says. "Our last names are
Brown, Black, and White, and one of us has brown hair,
another has black hair, and the third has white hair."

The person with brown hair says,
"And none of our hair color matches our name!"

"You are right," says Mrs. White.

If the man's hair isn't brown, what is the color
of Miss White's hair?

The man's hair isn't brown, and Mrs. Brown's hair isn't
brown. That means that Miss White's hair is brown.

Week Eighteen

SUNDAY | **GRAB BAG**

Assume these rules are true:

$$2 + 11 = 1$$
$$8 + 7 = 3$$
$$5 + 13 = 6$$

What is 3 + 18?

Turns out these rules *are* true, just not in base 10. Fooled you, right? Think about the other bases that you use regularly, and you'll find the answer with no problem.

In this case, you're working with base 12—in particular, time measured in a.m. and p.m.

So 2 a.m. + 11 hours = 1 p.m., and 8 a.m. + 7 hours = 3 p.m. Finally, 5 a.m. + 13 hours = 6 p.m. That means 3 a.m. + 18 hours is 9 p.m.

MONDAY | **NUMBER SENSE**

What is 24% of 73?

Finding percentages can be tricky, but this is not a tricky problem. Just remember that *of* means multiplication. With that in mind, you can multiply 24% and 73. But don't forget to change the percentage to a decimal: 24% = 0.24. Now multiply: 0.24 × 73 = 17.52.

There is another way to solve this problem, which helps with other percentage problems as well. Write a proportion—two ratios with an equal sign between them—and then solve for the unknown. When writing the proportion, you will know three of the four numbers. The percentage always goes with 100. That's because 24% is 24 out of 100. And the part always goes with the whole. In this situation, you can write the following proportion:

$$\frac{x}{73} = \frac{24}{100}$$

Again, notice that the part (x) is with the whole (73) and the percentage (24) is with 100. Also notice that the parts (x and 24) are in the numerators, while the wholes (73 and 100) are in the denominators. You can rearrange all of the numbers and still create a true proportion. Just be sure that you are keeping like pieces in the same locations.

Now you can cross multiply and solve for x. Cross multiplying, you'll get $100x = 1,752$. Divide each side of the equation by 100 to get $x = 17.52$. Whaddya know? That's the same answer from above.

Solve for x.

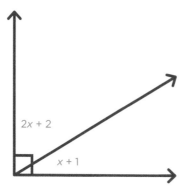

In this drawing, the two angles are *complementary*, which means that their measures add up to 90°. So to solve for x, you can add the expressions and set that equal to 90°: $2x + 2 + x + 1 = 90$.

To find x, combine like terms on the left side of the equation—the terms with x's and the terms without x's: $(2x + x) + (2 + 1) = 90 \rightarrow 3x + 3 = 90$.

Now subtract 3 from both sides of the equation: $3x = 87$. And finally, divide each side of the equation by 3: $x = 29$.

Week Nineteen

The figure below is called a triangular prism. The height of the triangular end is 4 units. The base of the triangular end is 6 units. The sides of the triangles are 5 units. The length of the rectangular piece is 10 units. What is the surface area of the figure? (Assume that you cannot see 2 rectangular faces and 1 triangular face.)

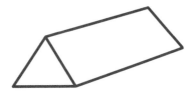

The formula for the area of a triangle is $\frac{1}{2}bh$.

Surface area is the area that covers the entire surface of the figure. If the figure were a pup tent, the surface area would be the amount of fabric needed to cover the tent—including the floor. Mark the dimensions you were given: height, base, and side of the triangle and length of the rectangle.

So to find the surface area, you'll just find the area of each of the faces—or the surfaces—of the figure. The key is to identify all of those faces. There are 2 triangles and 3 rectangles. The area of the triangles is $\frac{1}{2} \times 4 \times 6 = 12$ units^2. Since there are 2 triangles, you need to double that amount to find the area of both triangles: $2 \times 12 = 24$ units^2. To find the area of the rectangles, you need the length and the width of each rectangle. Then you can multiply: $10 \times 5 = 50$ units^2. There are 3 rectangles, so multiply: $3 \times 50 = 150$ units^2. Last step: just add the total area of the triangles and the total area of the rectangles: $24 + 150 = 174$ units^2. That's the surface area of the triangular prism.

To make rice, you add 1 cup of water per $3/4$ cup of rice. If you want to triple the recipe, how much rice will you need?

Tripling 1 cup of water is so, so easy, but that's not what you were asked to do. You need to triple $3/4$. Mathematically speaking, you're just multiplying: $3 \times 3/4 = 9/4 = 2\,1/4$ cups.

You can also think of this in terms of addition: $3/4 + 3/4 = 1\,1/2$ and $1\,1/2 + 3/4 = 2\,1/4$ cups.

Either way, that's a lot of rice.

FRIDAY | PROBABILITY & STATISTICS

The graph below shows the primary ways that people get their daily news. There were 3,429 people surveyed in all. How many people surveyed get their news from either newspapers or radio?

Primary Sources of News

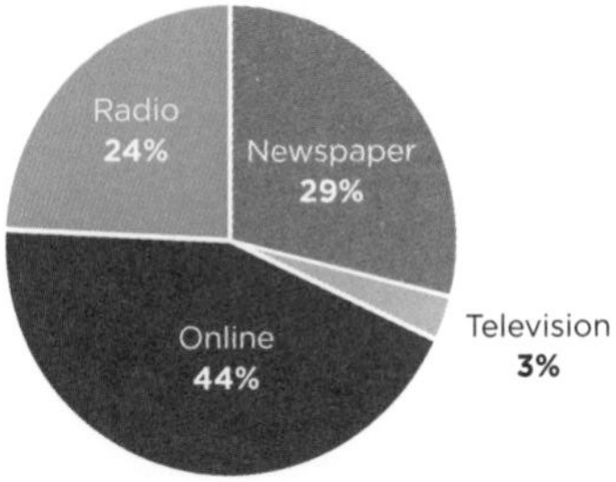

According to the graph, 29% of those surveyed primarily get their news from newspapers, while 24% turn to the radio. Since 3,429 people were surveyed, multiply 0.29 by 3,429 to find the total number of people who get their news from newspapers: 0.29 × 3,429 = 994. Do the same for the radio people: 0.24 × 3,429 = 823. Then add these to find out that 1,817 of the people surveyed typically open up the newspaper or turn to the radio for their daily news.

A phone manufacturing company has really screwed up the keypads for a batch of cell phones.

9	8	5
1	6	3
7	2	4
*	0	#

Only the bottom row is correct. The machines must be recalibrated to make sure that the numbers are in the correct order. But only one number switch can be made at a time. (For example, the 1 and 9 can be switched in one step.) What are the switches necessary to get the numbers on the keypads in the correct order? How can this be done in the fewest number of switches possible?

There are 5 switches in all. Exchange the 8 and the 2 to put both of these numbers in the correct places. Exchange the 3 and the 6, which puts the 6 in the correct place. Switch the 5 and 3, which puts both of these numbers in the correct places. Switch the 4 and 1, which puts the 4 in the correct place. And finally the 9 and 1 can be switched, which puts the keypad back to normal.

Did you make these switches in a different order? That's certainly possible. And it's also possible to make a couple of different exchanges than described above.

Week Nineteen

Use addition, subtraction, multiplication, and division with the following numbers to get 68. Do not rearrange the numbers.

2, 9, 8, 6

For this problem, you want to place +, −, ×, and / or ÷ between each of the numbers, so that the result is 68. You can approach this problem using trial and error, but it's a good idea to think things through first. Because 68 is much larger than the numbers in the list, you're likely going to multiply somewhere. For example, you could try 2 × 9 + 8 × 6. Keeping the order of operations in mind, this answer gives 18 + 48 or 66. Not quite. What would happen if you multiply the two middle numbers? That gives 72, which is too much. If you add 2, you get 74. Subtract 6 and you have the correct answer, 68. So one correct answer is 2 + 9 × 8 − 6.

Notice that the good old My Dear Aunt Sally (Multiply, Divide, Add, Subtract) is a big deal here. You can't simply add, multiply, and subtract from left to right. Instead, you need to multiply first and then perform the addition and subtraction (in either order).

Our number system is made up of overlapping categories. Whole numbers are the numbers from 0 to ∞ (infinity), excluding fractions, decimals, and irrational numbers. Rational numbers are positive and negative numbers that can be written as fractions. Irrational numbers are positive and negative numbers that cannot be written as fractions. (When written as decimals, irrational numbers never repeat and continue forever.) Integers are positive and negative whole numbers. And then there are imaginary numbers, which are the result of taking the square root of a negative number. Of course real numbers are not imaginary, but they do include both rational and irrational numbers.

Which of these categories fit inside other categories? In other words, which categories are part of a larger category?

The least intricate system of numbers is whole numbers, which are part of the integer system. Integers are also rational numbers. Rational numbers are not part of the irrational number system, but both rational and irrational numbers are real numbers. Sitting outside of the real number system are the imaginary numbers.

Week Twenty

Is there a solution to this problem? If so, what is it?

$$6 = \frac{x-1}{x}$$

Not all algebra problems have solutions—and for very good reasons. To find out if this one does, just try to solve for x. In this case, the first step is to multiply both sides of the equation by x. This will get the x out of the denominator on the right side: $6x = x - 1$. Now you can subtract x from both sides of the equation: $5x = -1$. Finally, you can divide each side of the equation by 5: $x = -\frac{1}{5}$.

It makes sense for x to equal $-\frac{1}{5}$, so that is the solution. But what if there were no solution? If you had found that $x = 0$, you'd be in no-solution land. That's because it's undefined to divide by zero.

What is the surface area of the cylinder below?
The height is 12 and the diameter is 4.

Think of unrolling the cylinder so that it's flat. That gives you 2 circles and 1 rectangle.

It's easy to find the area of the circles. Use the radius—half of the length of the diameter—in the formula: πr^2 or 3.14×2^2. This gives you 12.56. There are two circles, so double the area to find the total area of both circles: $12.56 \times 2 = 25.12$ square units.

Now for that big rectangle that you rolled out. The height of the rectangle is the height of the cylinder or 12. But what is the width? That's the distance around the circular bases, or the circumference, which is $2\pi r$. Substitute: $2 \times 3.14 \times 2 = 12.56$. (Huh. In this case, the area and the circumference of the circles are the same. That's because the square of 2 and doubling 2 both equal 4.) Now you have the width of the rectangle. The area is the width times the length: $12.56 \times 12 = 150.72$. There is only one rectangle, so you don't need to multiply the area of the rectangle by anything. Just add the areas of the circles with the area of the rectangle: $25.12 + 150.72 = 175.84$ square units.

THURSDAY | APPLICATION

Once again, you can't sleep. You have to get up at 5:30 a.m., and it's now 11:15 p.m. If you were to fall asleep right now, how many hours of sleep would you get?

How many of us have been in this place night after night? The trick to this problem is that the first time is p.m. and the second is a.m. This is all well and good, but hours are in base 12 and minutes are in base 60. In other words, the subtraction just ain't easy.

The simplest way to approach this problem is to picture a clock face. Imagine the minute and hour hands swinging forward from 11:15 p.m. to midnight. Forty-five minutes have passed, right? Next count the number of hours from midnight to 5:00 a.m. That would be five hours. Finally, there are 30 minutes from 5:00 a.m. to 5:30 a.m. Now mentally add these together, starting with the minutes: 45 minutes plus 30 minutes is 1 hour and 15 minutes. Add the 5 hours, and you get 6 hours and 15 minutes. Not enough sleep.

FRIDAY | **PROBABILITY & STATISTICS**

Sunshine Ice Cream shop has been doing a fine business at South Beach for generations. The new manager wants to track daily sales, comparing these numbers with the weather. He suspects that sales go up when the weather is nice. He notices that on sunny days, ice cream sales are indeed up, while on rainy days, sales are down. Can the manager conclude that sunny weather causes more customers to come in to buy ice cream?

This answer may seem perfectly obvious—but it's not. There's no way to determine without a shadow of a doubt that customers buy ice cream *because* the sun is shining. In fact, it is more likely that more people come to the beach in sunny weather, bringing a larger population of possible ice cream buyers.

This is a fine line, certainly, but it can make a big difference in how research is relayed. The most common example of this is with umbrella sales. Do more umbrella sales make it rain? Or does rain cause higher umbrella sales? Clearly, the latter is the more likely conclusion.

The bottom line is this: just because there is a correlation doesn't mean that causation exists. But a correlation can be enough to make this ice cream shop manager make some staffing changes. It doesn't really matter what brings more customers to the shop; on sunny days, he'd better have a bigger staff at the counter.

Four people are using a rowboat to get across a river. Julia gets along with everyone. Sam cannot get along with Eli, and Zoe cannot get along with Sam. Only three people fit in the boat at a time, and if two of them start fighting, Julia can't focus on rowing the boat. How can all four get across the river in the shortest number of trips without anyone fighting?

Since Julia gets along with everyone, she can row everyone across the river. Turns out Julia needs five trips across the river to get them all on the other side without fighting.

First she takes Eli and Zoe. (They get along just fine.) She leaves Zoe on the opposite bank, and then rows back to the original side with Eli. Then Eli gets out of the boat and Sam gets in. Julia rows Sam to the other side. Sam gets out of the boat, and Zoe gets in, leaving Julia and Zoe to row back to the original side. Julia picks up Eli, and the three row to the other side, where Sam is.

Week Twenty

SUNDAY | GRAB BAG

Clarke plants an ivy plant in the middle of a square garden. Every two days the ivy doubles in size. In 30 days the garden is completely covered with ivy. On what day was the garden half covered?

If you start at the beginning, you will get frustrated quickly. So why not start at the end? On day 30, the garden is completely covered. The ivy doubles in size every two days, so it was half covered on the twenty-eighth day.

MONDAY | NUMBER SENSE

$$4 \div 2 + 9 - 2 \times 1 = ?$$

If you approach this problem by moving from left to right, unfortunately you'll be going down the wrong mathematical road. In this case, you must follow the order of operations, which can be remembered with a simple phrase: Please Excuse My Dear Aunt Sally (parentheses, exponents, multiplication, division, addition, subtraction). There is division, addition, subtraction, and multiplication in this problem, so you'll only need that part of the mnemonic.

But here's the thing: multiplication and division can be done in any order, as can addition and subtraction. So you do have a little bit of wiggle room.

Here's one way to solve the problem:
$4 \div 2 + 9 - 2 \times 1 = 2 + 9 - 2 \times 1 = 2 + 9 - 2 = 11 - 2 = 9$.

But there are other ways to handle this, too. For example, when you get to 2 + 9 - 2, you can subtract 2 from 9 before adding 2. You'll get the same correct answer. Or you can subtract 2 from 2, leaving 9, the correct answer.

See? Math is flexible.

Week Twenty-One

TUESDAY | ALGEBRA

Is there a solution to this equation? If so, what is it?

$$x + 2 = x + 4$$

Not all equations are *true,* which means that not all equations have solutions. If you take a closer look, you can see that this equation doesn't make sense. It doesn't have a solution.

To be sure, attempt to solve the equation. First, subtract 2 from both sides of the equation: $x = x - 2$. Already, this is problematic. There is no value for x that makes the equation true. If you go forward to solve the equation, you need to subtract x from both sides of the equation. That leaves $0 = 2$, which you know is not a true statement.

Just like in life, not all problems have a solution. (Math can be a little philosophical, you know.)

The formula for the volume of a sphere is

$$\frac{4}{3}\pi r^3$$

What is the volume of a *semi-sphere* with diameter of 10?

There are two things to keep in mind here. First, the formula is for a sphere, and you are asked to find the volume of a semi-sphere. But that's simple to manage: find the volume of a sphere and then divide by 2.

Second, the formula requires that you have the radius. But you are given the diameter. So you need to find the radius before you can use the formula at all. Once again you're going to divide by 2, since the radius is half of the diameter.

Since the diameter of the semi-sphere is 10, the radius is 5. Use that value to substitute into the formula: $^4/_3 \times 3.14 \times (5)^3$. Now follow the order of operations. Evaluate the exponent first: $^4/_3 \times 3.14 \times 125$. Then multiply by 3.14: $^4/_3 \times 3.14 \times 125 = ^4/_3 \times 392.5$. The easiest way to handle the fraction is to divide by 3 and then multiply by 4. (Or you can do it the other way around—multiply by 4 and then divide by 3.) That gives you 523.33, if you round to the nearest hundredth.

But don't stop there. That's the volume of a sphere with a radius of 5. You were asked to find the volume of a semi-sphere with a radius of 5. So divide by 2 to get 261.67 (rounded to the nearest hundredth).

Week Twenty-One

THURSDAY | APPLICATION

A medication contains ¼ gram of active ingredient. If 1 gram equals 1,000 milligrams, how many milligrams of active ingredient are there in the medication?

Since each gram is equal to 1,000 milligrams and there is ¼ gram of active ingredient in the medication, there are 250 (1,000 × ¼ or 1,000 ÷ 4) milligrams of active ingredient in the medication.

Week Twenty-One

FRIDAY | PROBABILITY & STATISTICS

Nancy is vying for Courtney's job as a concert promoter. She wants to show that Courtney is not doing well and that she should be fired. Nancy has made a chart that shows the actual tickets sold over the last month compared to the goals set for that month. Has she demonstrated that Courtney is failing?

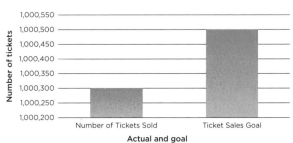

**Concert Ticket Sales:
Actual vs. Goal**

Those bars do look far apart. The ticket sales goal is much taller than the actual number of tickets sold. But if you look more closely, you can see that the scale of this graph is off. Instead of starting at 0, the graph starts at 1,000,200. And the values on the vertical axis are only 50 units apart. This creates an optical illusion; the difference between the two bars looks much larger than it actually is. In fact, Courtney is only 200 ticket sales away from her goal. And when you think about the number of tickets sold—1,000,300—that's not bad at all.

Nancy should look for someone else to pick on.

Week Twenty-One

SATURDAY | LOGIC

Dallas is Emory's father's nephew. Kai is Emory's cousin, but not Dallas's brother. Emory's father has one brother and no sisters. What is the relationship between Kai and Dallas?

Since Dallas is Emory's father's nephew, Dallas is male. Also, you know that Emory has two cousins: Dallas and Kai. Emory's father has one brother and no sisters, so all of Emory's cousins must be offspring of that brother. Kai is Emory's cousin, but is not a brother to Dallas. So Kai must be Dallas's sister.

SUNDAY | **GRAB BAG**

In a well-known Christmas carol, the singer is given a gift each day for 12 days. On the first day, he receives a partridge in a pear tree. On the second day, he receives 2 turtledoves *and another* partridge in a pear tree. On the third day, he receives 3 French hens, another pair of turtledoves, and another partridge in a pear tree. So each day, the singer receives a new gift, plus another set of all of the previous days' gifts. On the 12th day, he receives the following:

A partridge in a pear tree

Two turtledoves

Three French hens

Four calling birds

Five golden rings

Six geese a-laying

Seven swans a-swimming

Eight maids a-milking

Nine ladies dancing

Ten lords a-leaping

Eleven pipers piping

Twelve drummers drumming

How many gifts *in all* did the singer receive from his true love?

Week Twenty-One

SUNDAY | GRAB BAG

(continued from previous page)

It may make sense to simply add all of the gifts as they're listed above: 1 + 2 + 3 + 4 . . . But that only gives the gifts that the singer received on the *last* day. Remember, these gifts keep piling up each day.

Of course there are several ways to approach this problem. You can find the total number of each type of gift and add those together. For example, 1 partridge is given every day for 12 days. So there are 1 × 12 or 12 partridges. Two turtledoves are given each day for 11 days: 2 × 11 = 22. And so on. Then you can add the totals of each type of gift, which gives 364. (Not such a nice gift anymore, eh?)

Drawing a set of dots for each day's gift can help you come to this answer more quickly. (Draw the dots in triangles, rather than rows and columns.) And of course there's a fancy formula for finding the total number of gifts: $\frac{1}{6}(n)(n + 1)(n + 2)$, where n = 12. Plug it in and see for yourself—you'll get the correct answer.

MONDAY | **NUMBER SENSE**

-4, $^3/_4$, and -2.3 are all rational numbers.
π, $-3\sqrt{2}$, and $\sqrt{^3/_4}$ are irrational numbers.
Based on these examples, what is a rational number?
And what is an irrational number?

One of the basic structures of mathematics is the number system. To keep things organized, mathematicians have categorized numbers based on their differences and similarities. There are many different types of numbers, including integers (negative and positive whole numbers), rational numbers, irrational numbers, and real numbers (the set of rational and irrational numbers). There are even *complex numbers*, which include all real numbers, plus all imaginary numbers. (No kidding!)

If you look carefully at the numbers above, you may recognize similarities and differences. For example, rational and irrational numbers can be positive or negative. But there is a big difference that can be explained in two different ways. When a rational number is expressed as a decimal, the numbers to the right of the decimal point terminate or repeat in a predictable pattern. For example, $^3/_4$ can be written as 0.75. However, this is not true for the irrational numbers. Of course π is the most famous irrational number—it goes on forever without repeating. The same is true for many square roots, including the square root of 2 and the square root of $^3/_4$.

Another way of explaining this difference is by considering fractions. Rational numbers can be written as fractions, while irrational numbers cannot. (Even though $\sqrt{^3/_4}$ has a fraction in it, the number itself cannot be written as a fraction without a square root.)

TUESDAY | ALGEBRA

What does x^n mean? What is another way to describe it?

Math is full of fancy-schmancy symbols, like Σ and ! and ∞. But the little number to the right of a number or variable or expression is one of the easiest ones to remember. If the n were a 2, you'd know automatically that you need to square the x. (Whatever x is.) And *squaring* means multiplying the x by itself one time: $x^2 = x \times x$. If that little number is 3, you'd multiply the x by itself 2 times: $x^3 = x \times x \times x$.

So x^n means to multiply x by itself $n - 1$ times. Wait, where did the $n - 1$ come from? It came from the examples above. The number of multiplication symbols is the same as the number of times you multiply, and $x^2 = x \times x$, while $x^3 = x \times x \times x$. Therefore, $x^n = x \times x \times x \times \ldots \times x$, where the number of multiplication symbols is $n - 1$.

It's true. Sometimes explaining the math is more complicated than the math itself.

Week Twenty-Two

WEDNESDAY | **GEOMETRY**

A semi-sphere has a diameter of 10 meters. What is the volume? (The formula for the volume of a sphere is $\frac{4}{3}\pi r^3$.)

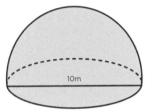

10m

A semi-sphere has half of the volume of a sphere. So if you find the volume of a sphere with a diameter of 10 meters, you can take half of this figure to find the volume of the semi-sphere.

The formula for the volume of a sphere requires the radius of the sphere. But you have the diameter. Good thing that the diameter is twice the length of the radius. The radius of this semi-sphere is 5 meters. Now just substitute 5 for r and 3.14 for π and simplify:

$V = \frac{4}{3}(3.14)(5)^3 = 1.33(3.14)(125) = 522.025$.

(Sometimes it's helpful to change fractions to decimals. Since π is being approximated anyway, it's fine to convert $\frac{4}{3}$ to 1.33.)

But remember, this is the volume of a sphere with a diameter of 10. You still need to take half of this value to find the volume of the semi-sphere: $522.025 \div 2 = 261.0125$ m^3. And that's all there is to it.

The house that Tasha wants to purchase costs $213,000. She needs to make a 15% down payment. How much is the down payment?

This is a simple percentage problem. What is 15% of $213,000? To find out, change 15% to a decimal and then multiply: 0.15 × 213,000 = $31,950.

But what if you didn't remember how to solve this percentage problem? You can always set up a proportion or a pair of equal fractions. The percentage is always in one fraction: $^{15}/_{100}$. The other fraction is the part over the whole. In this case, that means the down payment over the total cost of the house. You don't know the down payment, so that's where your variable goes:

$$\frac{15}{100} = \frac{x}{213,000}$$

Now you can simply cross multiply and solve for x: 15 × 213,000 = 100x → 3,195,000 = 100x → 31,950. And that's a chunk of cash.

Week Twenty-Two

FRIDAY | PROBABILITY & STATISTICS

The Red Roses, Red Roses Floral Shop has tracked its Valentine's Day sales for five years. Using this data, the shop can plan for the current year's holiday, making sure it has enough roses in stock. Based on this data, how many roses should the shop purchase this year?

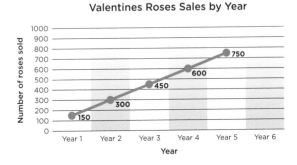

Red Roses, Red Roses Floral Shop Valentines Roses Sales by Year

In this number line, it's easy to see the amount of roses sold each year. In year 1, the store sold 150 roses; in year 2, it sold 300, and so on.

Each year, the sales increase by 150 roses. So by year 6, the shop should order 150 more roses than the previous year. That's 750 + 150 or 900 roses.

Week Twenty-Two

SATURDAY | LOGIC

A red maple tree, a black walnut tree, and a white ash tree grow side by side in the woods. The red maple is home to a blue jay, the black walnut has a cardinal's nest, and the white ash has a crow's nest. The tree next to the white ash has one knot. The tree with a black bird has two knots. Either the tree with the blue bird or the tree with the black bird has three knots. The white ash and black walnut trees are not next to each other. Which tree has one knot?

Whew, that's a ton of information. A table or list will probably help out. What you have are three characteristics of each tree: the type of tree, the type of bird living in the tree, and the number of knots in the tree. There is also some information about the order of the trees.

Already, you know which tree has which bird. The red maple has the blue jay, the black walnut has the cardinal, and the white ash has the crow. It's helpful to know that a blue jay is blue, a cardinal is red, and a crow is black. Now you can take the remainder of the information to find the number of knots in each tree. The tree with the black bird—the white ash—has two knots. Also you know that either the tree with the blue bird—the red maple—or the tree with the black bird—the white ash—has three knots. Well, you just found out that the white ash has two knots, so the red maple must have three knots. That leaves the black walnut with one knot.

Did you notice that the order of the trees really didn't matter? That's called a *red herring*.

Week Twenty-Two

SUNDAY | GRAB BAG

Seven sisters have 7 cats each. Each of the cats has its own litter of 7 kittens. How many legs do all of these people and felines have?

You can approach this problem several different ways: by finding the number of people and cats first and then the number of legs or by finding the total number of people legs, cat legs, and kitten legs.

Since there are 7 people who have 2 legs each, there are 14 people legs. But how many cats are there? Each sister has 7 cats, so there are 7 × 7 or 49 cats. Each of these cats has 4 legs, so there are 49 × 4 or 196 cat legs. Now for the kittens. Each cat has 7 kittens. There are 49 cats, so there are 49 × 7 or 343 kittens. Each kitten has 4 legs, so there are 343 × 4 or 1,372 kitten legs.

All that's left to do is to add the number of people legs, cat legs, and kitten legs: 14 + 196 + 1,372 = 1,582. (Perhaps these sisters should consider spaying.)

Week Twenty-Three

MONDAY | NUMBER SENSE

The Least Common Multiple (LCM) of two numbers is the smallest number that each will divide into evenly. What is the LCM of 9 and 12?

Here's another way to describe a *multiple*: It's created when you multiply a number by a whole number. So the LCM is the largest multiple that two numbers have in common. The two numbers above are relatively small, so listing the multiples is no biggie. Start with 9, multiplying by 1, 2, 3, 4, etc.: 9, 18, 27, 36, 45, 54, and so on. Now list the first six or so multiples of 12: 12, 24, 36, 48, 60, 72. Choose the smallest number that appears in both lists, 36, which is the LCM for both of these numbers.

So why is LCM a big deal? Fractions. When you add or subtract fractions that do not have the same number in the denominator, you have to first find a common denominator. It doesn't have to be the LCM of both numbers, but using the LCM can keep things a little simpler in the long run.

For example, $\frac{1}{9} + \frac{1}{12} = \frac{4}{36} + \frac{3}{36} = \frac{7}{36}$. All done thanks to LCM.

If $x = 4$, what is $2^x + 9$?

In this expression, the variable is in an exponent. What the heck does that mean? Just substitute, and see what happens: $2^4 + 9$. So the process is pretty darned simple. Start by multiplying 2 by itself 3 times: $2 \times 2 \times 2 \times 2 = 16$. Now add 9 to get 25.

Sometimes problems look harder than they actually are.

WEDNESDAY | GEOMETRY

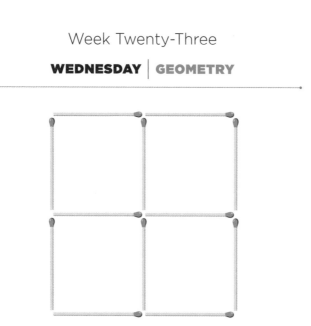

Remove two matchsticks to create two squares.

If you're messing around with the matchsticks around the outside, you're barking up the wrong tree. Instead, you need to work with the ones on the inside. Another tip? The two squares will not be the same size. Remove any of the matchsticks that are perpendicular to one another and on the inside of the figure, and you'll have one large square and one small square inside it.

THURSDAY | APPLICATION

Forty percent of the 60 tulips in your garden are yellow. How many tulips are *not* yellow?

This is not your typical percentage problem. You can't just multiply these two values together and get the answer. Instead, you need to think about the question that's being asked. If 40% of the tulips are yellow, 60% are *not* yellow. *Now* you can multiply: 0.60 × 60 = 36. So 36 of the tulips are not yellow.

Want another approach? Find out how many yellow tulips there are (0.40 × 60 = 24). Then subtract from the total number of tulips (60 – 24 = 36). Same answer, different process.

Week Twenty-Three

There are two cats—one black and one white. What is the probability that both cats are male? If you know that the black cat is male, what is the probability that the white cat is female?

In terms of gender, there are four possibilities: (female, female), (male, male), (female, male), (male, female). It makes no difference which cat is black and which is white. There is a 1 in 4 chance that both cats are male.

But you know that the black cat is male. The other cat is definitely white, and it is either female or male. That's one positive outcome (male) out of two possible outcomes (male or female). So the chance that the white cat is male, given that the black cat is female? One-half.

SATURDAY | LOGIC

Two babies, Harper and Hayden, are chatting in the hospital nursery. Harper says, "I'm a girl." Hayden says, "I'm a boy." If one of them is a girl and one is a boy, and if at least one of them is lying, who is the boy and who is the girl?

Harper is a boy, and Hayden is a girl. In other words, they are both lying. If only one of them were lying, they would both be girls or both boys.

Week Twenty-Three

SUNDAY | GRAB BAG

Leonhard Euler, a Swiss mathematician,
came up with this problem.

A river contains 2 islands. There are 7 bridges that connect
the banks of the river and the islands. Can a person begin at
one bridge and then cross each bridge exactly once? If not,
what can be done to the bridges to make this process work?

You may have already guessed that 7 bridges cannot
be crossed exactly once. However, if you remove one
of the bridges—the one that connects the 2 islands—you
can cross each of the bridges exactly once.

Why? Great question. This has to do with what
Euler calls odd vertices. (Hang in there; it's not that
complicated.) If you put 1 dot on each of the landmasses,
you'd have 4 vertices. One vertex is on the island on the
left and 1 is on the island on the right. Then put two more
vertices above and below the point on the left island.
One vertex (the one on the island on the left) has 5 paths
or bridges to it. The vertices above, below, and to the
right have 3 paths (bridges) to them. Euler found that
if a figure has more than 2 vertices with an odd number
of paths to them, there is no one path that crosses each
individual path exactly once.

Week Twenty-Four

MONDAY | **NUMBER SENSE**

$$-2 \times 4 = ?$$

Ack! If that 2 were positive, this would be a really easy problem. And if you remember the rules for multiplying and dividing positive and negative numbers, you'd be just fine. But say you don't. Just like with adding and subtracting integers, a number line can help.

Draw the number line, including all of the integers from -10 to 10. That means the 0 is going to be smack-dab in the middle.

Starting at 0, you're going to make 4 jumps of -2. What the heck does that mean? Well, you'll move to the left 2 units, and repeat this for a total of 4 times. On the first jump, you'll land on -2. The second jump puts you at -4, and the third jump gets you to -6. On the last jump, you land on the answer: -8.

But what about that rule? If you can remember it, this rule is much easier to follow than using a number line. Whenever you multiply (or divide) two numbers with opposite signs, the answer will always be negative. In this case, multiply 2 and 4 to get 8. Then make the 8 negative, since you are multiplying a negative (-2) by a positive (4).

If $x = \frac{1}{8}$, what is $16x + 3$?

U h-oh, there's a fraction in there. But you shouldn't be afraid of it, especially in this problem.

As with other problems like this, start by substituting: $16(\frac{1}{8}) + 3$. The order of operations says that you need to multiply before adding, but what is $16 \times \frac{1}{8}$? One way to think of this is as division: $16 \div 8 = 2$. Then add 3 to get 5, and you're all done.

Of course you can use a calculator, but why, when your powerful brain can handle it on its own?

Week Twenty-Four

WEDNESDAY | GEOMETRY

To create the orange figure, has the blue figure been translated (slid), rotated, or reflected? Explain your answer.

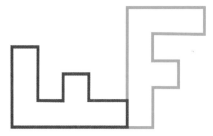

Notice that the orientation of the figures is different. In the blue figure, the longest side is horizontal. In the orange figure, the longest side is vertical. That means a rotation was performed. In fact, the rotation was 90° around the bottom right point of the blue figure.

THURSDAY | **APPLICATION**

You've had a wonderful meal at a fine restaurant. Your bill totals $134.85. The service was so good that you want to leave a 20% tip. About how much is the tip? (To really exercise your brain, don't use a calculator.)

You can certainly multiply the total by 0.20 (which is the same as 20%), but that's a tough thing to do in your head. Instead, try taking 10% of the total—by moving the decimal point one place to the left. That gives you 13.485, which rounds to 13.50. (Rounding to the nearest half-dollar makes the problem simpler.) Now just double that amount to get $27. This isn't the exact amount of the tip, but it's close enough.

Week Twenty-Four

FRIDAY | PROBABILITY & STATISTICS

This graph shows the cost of a gallon of milk in five stores over three years. Is this a good representation of the data? If not, what should be done?

Cost of Milk in Five Stores, Over Three Years

This graph is a bit confusing. First of all, there is no key that shows which year is which. Secondly, the vertical axis is not detailed enough to show the minor differences in the price of a gallon of milk. (And it doesn't have units!) Also, the pyramids are a poor choice to demonstrate the differences in the cost of milk over the years. If you think about the volume of the pyramids, the milk in the orange year would cost much less than the milk in the blue year. But is that what was intended?

A line graph would probably be a much better idea. Each year would have one line, showing how the cost increased and decreased over time. With that display, it is easier to compare yearly costs in each store and compare the trends by store. But a key is necessary to show which line demonstrates which year.

SATURDAY | LOGIC

The town drunk is seen one night on the corner shouting, "Brothers and sisters I have none, but that man's father is my father's son." Who on earth is he talking about?

First off, you know that the drunk has no brothers or sisters, so a sibling is out of the question. In fact, the person must be male, because the drunk says "that man." Now break down the phrase "that man's father is my father's son." Think of this chronologically. The unknown man's father is the drunk's father's son. That means that the drunk is the unknown man's father, and so the unknown man is the drunk's son.

SUNDAY | GRAB BAG

Two six-sided dice are sitting on a counter. The sum of the dots on visible sides is 32. Which faces of the dice are *not* showing?

You could think about all of the numbers between 1 and 6 that add up to 32. Or you could *reverse* this thinking. If the dots showing have a sum of 32, what is the sum of the dots that are not showing? Breaking that number down will give you the dots that aren't showing.

First add up the dots on two dice:
1 + 2 + 3 + 4 + 5 + 6 + 1 + 2 + 3 + 4 + 5 + 6 = 36.
Now, find the sum of the dots that are not showing:
36 – 32 = 4. Since there are two dice, there are two faces not showing. What two numbers add up to 4? That would be 1 and 3—and 2 and 2.

Sometimes looking at a problem backward makes it easier to solve.

MONDAY | **NUMBER SENSE**

$$-5 \times -2 = ?$$

Another one of those multiplying integers problems, except this time you're asked to multiply two negative numbers. Is the answer going to be positive or negative? If you remember the rule, you know. And if you don't, just pull out that number line.

Start at 0. This time, you'll take -2 jumps of -5 numbers. So instead of going to the left 2 jumps of 5 numbers, you'll actually go to the *right* 2 jumps of 5 numbers. The first jump gets you to 5, and the second gets you to 10, which is the correct answer.

That can be confusing, so it may be more helpful to remember the rule in this case. Whenever you multiply two negative numbers, the answer will be positive. Always and forever. So in this case, multiply the numbers without considering their signs: $5 \times 2 = 10$. The answer is positive because 5 and 2 are both negative.

If $x = 2$, which of the following is the largest expression?

$$x(x - 2)(x + 1)$$
$$(x + 1)(x + 2)(x + 3)$$
$$18x + 12$$
$$x(x + 12)$$

Perhaps the most obvious way to answer the question is by evaluating each of the expressions for $x = 2$. That means substituting 2 wherever you see x, and then finding the value of the expressions.

In the first expression, substitute 2 for x in three places: $2(2 - 2)(2 + 1)$. That gives $2 \times 0 \times 3$, and since anything times 0 is 0, the value is 0. It's not likely that this is going to be the largest value.

In the second expression, the substitution goes like this: $(2 + 1)(2 + 2)(2 + 3)$. Add first, because of the parentheses, to get $3 \times 4 \times 5 = 60$. That's a pretty big number, so you might have a winner.

But first, check the third and fourth expressions: $18 \times 2 + 12 = 36 + 12 = 48$, and $2(2 + 12) = 2 \times 14 = 28$. Clearly the second expression is the largest, when $x = 2$.

WEDNESDAY | GEOMETRY

To create the orange figure, has the blue figure been translated (slid), rotated, or reflected? Explain your answer.

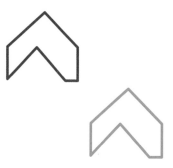

For these figures, the orientation has not changed. In other words, the vertex at the top of the blue figure is in the same place as the vertex on the top of the orange figure. This means that the blue figure was *translated*—slid down and to the right—to create the orange figure.

Week Twenty-Five

THURSDAY | APPLICATION

Last year your commute was 15 minutes long. This year, it is 40% longer. How long is your commute this year?

Like most problems, this is not so cut and dry. You're not being asked 40% of 15 and that's it. Instead, you need to then add 40% of 15 to 15.

Are you following me? Let's break it down. First start by finding 40% of 15: $0.40 \times 15 = 6$. So right now you know that your commute is 6 minutes longer. That means that your commute is 15 + 6 or 21 minutes.

FRIDAY | **PROBABILITY & STATISTICS**

This scatter plot shows the relationship between the amount of time an individual spent training for a 5K and that person's time in the race. What can you say about this relationship? What can you not say about this relationship?

Training Weeks and 5K Time

otice how the dots start high and then get lower, as the data goes from left to right? This is called a negative correlation, but there are other, less technical ways to describe it. In short, this data shows that the longer someone trains for a 5K, the shorter the person's time will be. Typically. Notice that there is one outlier—one person trained for 10 weeks, but took more than 40 minutes to finish the race. This is proof of the next conclusion. While there is a correlation between the two pieces, this graph does not show *causation*. In fact, it is very difficult to show that training longer will *cause* a person to run faster. So many other elements are at play here, including how hard the person trains, whether that individual felt well on the day of the race, and so forth. So while training time can influence a person's 5K time, it's not the only element that will.

SATURDAY | LOGIC

Assume the following are true:
Many children are boys. All boys are rowdy.

Which statement is true?

1. Some children are rowdy.
2. No boys are children.

It's tempting to say that the last statement is automatically false. Of course boys are children. But in logic problems like this one, the context doesn't matter, just the logical statements.

So look at the first two statements. Many children are boys, and all boys are rowdy. (See? You know in real life that not all boys are rowdy, but because you need to accept the first statements as fact, you've got to roll with it.) Now look at the first conclusion: some children are rowdy. Since many children are boys and all boys are rowdy, some children are rowdy. So the first conclusion is true.

Now the second conclusion: no boys are children. This contradicts the first given statement that many children are boys. If many children are boys, then at least some boys are children. Therefore, the second conclusion is false.

Week Twenty-Five

SUNDAY | **GRAB BAG**

1. A number has 3 digits and is odd.
2. Two digits are the same.
3. The sum of the digits in the tens and ones places is odd.
4. The sum of the digits is 4.

What is the number?

Start with the last clue. The sum of the 3 digits is 4. So the digits can be 1, 1, and 2; 0, 1, and 3; 0, 0, and 4; or 0, 2, and 2. Two of the digits are the same, leaving 1, 1, and 2 or 0, 2, and 2 as the digits. But the number is odd, so the digits must be 1, 1, and 2.

But in what order? Since the number is odd, the ones digit will be 1. Since the sum of the tens and ones place is odd, the tens digit is 2. That leaves 1 in the hundreds place. And 121 is the mystery number.

Week Twenty-Six

MONDAY | NUMBER SENSE

The Greatest Common Factor (GCF) is the largest integer that will divide evenly into a set of integers. What is the GCF of 24 and 18?

The GCF is the little sister of LCM. But instead of multiples, you're looking for factors. *Factor* is another way of describing an integer (a positive or negative whole number) that divides evenly into another integer. So one way to approach this problem is to list all of the factors of each of these numbers. The factors of 24 are 1, 2, 3, 4, 6, 8, 12, and 24. (These are arranged in order from smallest to largest, which is helpful. But you can certainly list them as pairs of factors: 1, 24, 2, 12, 3, 8, 4, 6.)

Now list all of the factors of 18: 1, 2, 3, 6, 9, 18. What is the largest number that appears in both lists? That would be 6, so that's the GCF.

But why in the world is it necessary to find a GCF? The GCF is helpful when simplifying fractions. For example, $^{18}/_{24}$ can be simplified to $^3/_4$ by dividing the top and bottom by 6, the GCF of both numbers.

Solve for x.

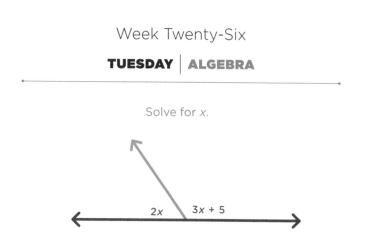

The two angles in the picture are *supplementary*, which means that they add up to 180°. You know that for sure because the two angles form a straight line. Find x by adding the two algebraic expressions and setting the sum equal to 180: $2x + 3x + 5 = 180$.

Now you can combine like terms in the equation. Add the terms with x's: $5x + 5 = 180$. Subtract 5 from each side of the equation: $5x = 175$. Then divide each side of the equation by 5 to get $x = 35$. So with just one little geometry fact and a few steps of algebra, the problem is simple Simon.

In this image, 4 matchsticks enclose a space that has an area of 1 square matchstick.

In this image, 6 matchsticks enclose a space that has an area of 2 square matchsticks.

Using 12 matchsticks, create a figure with an area of 5 square matchsticks. (None of the matchsticks can overlap or cross.)

You won't make a rectangle out of the matchsticks. Instead, create a letter *L*, using three matchsticks along the vertical side and the horizontal side. The remaining matchsticks make up the *L*.

Here's how. Take 3 matchsticks and line them up vertically. At the bottom of this column, line up 3 matchsticks horizontally going to the right. (And line up the first horizontal matchstick with the bottom matchstick of the vertical column.)

(continued from previous page)

At the far right matchstick on the horizontal row, add 1 vertical matchstick. Working counterclockwise and always adding matchsticks to the ends of matchsticks, add 2 more, horizontally, and going to the left. Then add 2 more, vertically and going up. Finally, close off the figure with 1 horizontal matchstick going to the left. The figure you end up with looks like the letter L. It has 12 matchsticks, none of which overlap or cross. But what about the area?

Here's one way to confirm the area. Find the area of each of the rectangles and add. Regardless of how you divide the figure, you'll get 2 small rectangles that have areas of 2 and 3. Add those together and you get 5.

Or you can draw lines to fill out the figure into a square. The square will have an area of 3 × 3 or 9.

Subtract the area of the smaller square, which was added. That square has an area of 4, and 9 – 4 = 5. Two approaches, one answer.

THURSDAY | APPLICATION

A pizza place has 5 tables that seat 6 people each. A hungry baseball team comes in and takes up all of the seats. They want pizza, of course, and each pie is cut into 8 slices. What is the fewest number of pizzas the team should purchase so that each table gets at least one whole pizza with no slices left over?

Since each table has 6 people and each pizza has 8 slices, you're looking for the least common multiple (LCM) of 6 and 8—the smallest number that both 6 and 8 divide into evenly. If that answer doesn't come to mind, start by looking at the multiples of 6: 6, 12, 18, 24, 30, 36, . . .

Are any of these also multiples of 8? If so, what is the smallest? Clearly, the team needs to order 24 pizzas. These won't be evenly divided among all of the 5 tables—one table will get 4 pizzas, while the other 4 will get 5 pizzas each—but each table will get whole pizzas.

Week Twenty-Six

FRIDAY | PROBABILITY & STATISTICS

Penelope Peters-Peterson is an unethical residential housing developer. She advertises her new homes in a variety of places. She doesn't have much room, and price is the most important aspect for buyers. Of course, Penelope wants to put the best price out there. In one development, homes have the following prices: $100,000, $600,000, $650,000, $650,000, $700,000, and $750,000. Which statistic—mean or median—do you think Penelope will choose to advertise?

Before making this decision, it might help to find the median and mean, so that you can compare the two. The median is the middle number, when the values are arranged in order from smallest to largest (or largest to smallest). There are two middle numbers in this list, and since they are both $650,000, that's the median price. The mean is the sum of the values divided by the number of values. Add these values, and you get 3,450,000. Divide by 6 to get $575,000.

The mean and median are different, but since the mean is smaller, that's the one Penelope will choose. But the mean is not the most representative number. (Remember, Penelope isn't honest.) Notice that the lowest home price is *well* below the next lowest home price. This is called an *outlier*. That $100,000 price tag is so different from the others that it brings the mean down. When you have a set of data like this, it's more ethical to use the median.

Week Twenty-Six

SATURDAY | LOGIC

The big horse race has just finished. You don't have the ranking of the race, but you can figure it out with these clues:

1. Money Maker finished before Shoo Fly, but after Momma's Sunday Hat.

2. Lickety Split finished before Momma's Sunday Hat.

3. King George and Hey Good Lookin' finished before Money Maker.

4. King George finished before Hey Good Lookin' and after Momma's Sunday Hat.

In what order did the horses finish the race?

Money Maker is between Shoo Fly and Momma's Sunday Hat, which is so far in the lead. Lickety Split is before Momma's Sunday Hat, so the order is now Lickety Split, Momma's Sunday Hat, Money Maker, and Shoo Fly. But what about King George and Hey Good Lookin'? These two horses finished before Money Maker, and King George finished before Hey Good Lookin' and after Momma's Sunday Hat, in that order.

So the final order is Lickety Split, Momma's Sunday Hat, King George, Hey Good Lookin', Money Maker, and Shoo Fly.

Whoa! Hopefully you kept up better than that last horse did.

SUNDAY | GRAB BAG

What is the sum of the first 50 odd numbers?

That's a lot of numbers. Listing and then adding them up is a pretty big process. So why not look for a pattern? The first 3 odd numbers are 1, 3, 5. Add those and you get 9. Interestingly, 3 × 3 = 9. But is this just a coincidence?

Look at the first 7 odd numbers: 1, 3, 5, 7, 9, 11, 13. Add these together and you get 49. And 7 × 7 = 49. Aha! Turns out this pattern works with any odd numbers that you add together.

So you can apply this pattern to the first 50 odd numbers. Since there are 50 of them, you can simply multiply: 50 × 50 = 2,500. Feel free to take the long way around; there's no shame in that. But this is a pretty nifty shortcut.

What is $\frac{1}{4} + \frac{3}{10}$?

Oh, fractions. How they vex. Or not. If you remember your elementary math classes, this problem is a cinch. Looking closely at these numbers, you can see that the issue is in the denominators. To add or subtraction fractions, the denominators must be the same. The answer will be the sum of the numerators over the common denominator.

(You could just change each of these to a decimal and then add. But adding fractions is a tougher cranial workout.)

So how can you make the different denominators the same? The good old least common multiple or LCM. This is the smallest number that both 4 and 10 divide into evenly. (Do you need the smallest number? No, but it does help.) Both 4 and 10 divide evenly into 40, but there is a smaller number that works: 20.

Just putting a 20 in each denominator isn't going to do it. You also need to deal with the numerators. And here's where you'll use something called the multiplicative identity. When you multiply any number by 1, the result is the original number. And in fractional form, 1 is any number over itself. So how would you multiply by 4 to get 20? That would be 5, right? So multiply $\frac{1}{4}$ by $\frac{5}{5}$. To multiply two fractions, just multiply the denominators and multiply the numerators. So $\frac{1}{4} \times \frac{5}{5} = \frac{5}{20}$. Do the same with the second fraction, this time multiplying by $\frac{2}{2}$: $\frac{3}{10} \times \frac{2}{2} = \frac{6}{20}$

Finally, you can add: $\frac{5}{20} + \frac{6}{20} = \frac{11}{20}$. And that's your answer. Holy moly, that's a lot of work for addition.

Week Twenty-Seven

TUESDAY | ALGEBRA

What is another way to write this expression?

$$3(x + 12)$$

This problem requires the distributive property. You are multiplying a single term by a *binomial expression* (meaning an expression with two terms). To do this, you must distribute the single term to each of the terms in the binomial. In other words, multiply the 3 and the x, and then multiply the 3 and the 12. Leave the addition alone. When you distribute, you get: $3x + 3 \times 12$, which is $3x + 36$.

WEDNESDAY | GEOMETRY

To create the orange figure, has the blue figure
been translated (slid), rotated, or reflected?
Explain your answer.

The orientation of the blue figure has changed to
create the orange figure. So you can forget about a
translation (slide), which keeps the orientation. And while
a rotation could be possible, the bottom angle of the blue
figure sticks out more than the top angle, so a rotation
doesn't make sense either. These figures are mirror
images, so the blue has been reflected over a vertical line
to create the orange.

You have 5½ yards of fabric to make napkins. The fabric is 45 inches wide. How many napkins can you make if each napkin is 16 inches × 16 inches and has a ½-inch hem?

Okay, to solve this problem you need to understand that the hem of a napkin goes all the way around the napkin. So when cutting out the fabric for one napkin, you'll add ½ inch per side of the napkin, which is 1 inch per dimension. In other words, you'll cut out squares that are 17 inches × 17 inches.

How many of these will fit along the width of the fabric? Since the fabric is 45 inches wide, you can fit only 2 napkins along the width. That's because 45 ÷ 17 = 2.65. You can't have 0.65 of a napkin, so you'll round down to 2.

Next, figure out the number of napkins that can fit along the length of the fabric. The fabric length is 5½ yards. It's probably easiest to convert this to inches. There are 36 inches in a yard: 5.5 × 36 = 198 inches. Divide by 17, to see how many napkins fit along the length of the fabric: 11.65. Again, you can't have 0.65 of a napkin, so 11 of these napkins fit along the length.

Now you've figured out that 2 napkins fit along the width and 11 fit along the length. That's 22 napkins in all —with fabric left over.

Louisa is fired up about the school board race. She decides to go down to the local supermarket and ask people whom they're going to vote for. After two hours, she takes her results and leaves. Can you tell if Louisa has good data? Why or why not?

A good poll will reflect the opinions of the larger population, but there are some really important rules to follow. The first is that the sample must be completely random. While Louisa has asked total strangers, she hasn't created a random sample.

She was at a particular supermarket at a particular time on a particular day. The location probably eliminates a large proportion of the population who shop elsewhere. The time of day also matters. If Louisa is there between 9 and 5, she is missing people who work during the day. On the other hand, if she's there in the evening, she's missing people who do night-shift work. Lastly, a good sample will have at least 1,000 responses. It's not likely that Louisa will get that in two hours at a grocery store.

To get an accurate poll, Louisa needs to select her sample more carefully.

The county bake-off works like an elimination tournament. The entries are divided into pairs (with one left over, if there are an odd number of entries), and these pairs are judged. The losing baker in the pair is eliminated.

If there are 38 bakers in the contest, how many bake-offs must occur before the champion is crowned?

Don't think too hard about this one. It's not as difficult as you think.

Since there are 38 entries and only one winner, 37 people will lose once. That means there are 37 bake-offs.

SUNDAY | **GRAB BAG**

1. A number has 4 unique digits and is odd.
2. The sum of the digits is 10.
3. Each digit is greater than zero.
4. The largest digit is in the thousandths place.
5. The digits in the tens and ones places have a sum of 3.

What is this number?

There's a lot going on here. Start with the last statement: the sum of the tens and ones place is 3. There are only two options here, since none of the digits can be zero: 1 + 2 or 2 + 1. To get these in order, look at the first statement: the last digit must be odd. This means the last two digits are 2 and 1.

Now for the remaining digits. The second statement is helpful here. Since all of the digits add up to 10, find the sum of the first two digits by subtracting 3 from 10. The first two digits have a sum of 7. It's pretty easy to list the possibilities here: 1 + 6, 2 + 5, 3 + 4, 4 + 3, 5 + 2, 6 + 1. The first, second, fifth, and sixth are not possible, since each of the numbers can only be used once. (Both 2 and 1 are already in the tens and ones places.) So you're left with 3 and 4. What order should they be in? Put the 4 first, since the largest digit is in the thousands place. That means 4,321 is the mystery number.

MONDAY | **NUMBER SENSE**

What is $-\frac{1}{6} + \frac{7}{9}$?

More fractions. You're adding again, but this time, you have a negative fraction plus a positive one. First, take a look at the fractions.

The denominators are different, so you need to find a common denominator, or least common multiple (LCM) of 6 and 9. The smallest number that both 6 and 9 divide into evenly is 18. Now you can rewrite the fractions with that denominator by multiplying each fraction by a number over itself.

In the first fraction, you'll multiply by $\frac{3}{3}$, because $3 \times 6 = 18$. In the second fraction, you'll multiply by $\frac{2}{2}$, because $2 \times 9 = 18$. Make sure to keep that negative sign in place: $-\frac{1}{6} \times \frac{3}{3} = -\frac{3}{18}$ and $\frac{7}{9} \times \frac{2}{2} = \frac{14}{18}$.

Now add: $-\frac{3}{18} + \frac{14}{18}$. You'll have to remember how add a negative and a positive number. Just find the difference between the two numerators ($14 - 3$). The answer will be positive, because (without the sign) the positive fraction is larger than the negative one: $\frac{11}{18}$.

If $x = 0$, evaluate

$$\frac{x+1}{x}$$

Did you notice that this was a trick question? Dividing by zero is undefined. Or if you look at it another way, you can't have a zero in the denominator of a fraction. So there is no defined answer for this problem. Really and truly.

But just to confirm, substitute and evaluate: $(0 + 1)/0 = 1/0$. Since you can't divide by 0, the answer is indeed undefined.

The Pythagorean theorem says that the sum of the squares of two legs of a right triangle is equal to the square of the hypotenuse:

$$a^2 + b^2 = c^2$$

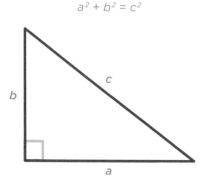

(The hypotenuse is the side opposite the right angle, and the legs are the sides that make up the right angle.)

If $a = 4$ and $b = 3$, what is c?

So while this is technically a geometry problem, you're going to use some algebra and basic arithmetic to find out what c is. First substitute what you know into the formula: $(4)^2 + (3)^2 = c^2$.

Now solve for c: $16 + 9 = c^2 \rightarrow 25 = c^2 \rightarrow 5 = c$. Strictly speaking, the answer is $c = \pm5$, but since length cannot be negative, you'll only take the positive value.

The numbers 3, 4, and 5 are special—called a *Pythagorean triple*. There aren't many whole numbers that fit this very famous theorem. In fact, there are no Pythagorean triples that contain a 6 or an 8.

Week Twenty-Eight

THURSDAY | APPLICATION

Hot dogs come in packages of 10. Hot dog buns come in packages of 8. What is the smallest number of packages of hot dogs and hot dog buns you should purchase to have exactly one hot dog per hot dog bun?

Who thought it was a good idea to package hot dogs and hot dog buns this way? At any rate, it's up to us consumers to figure out how many packages of each to buy. And this is where the least common multiple (LCM) comes in handy. You want to know the smallest number that both 10 and 8 divide into evenly. That will give you the smallest number of hot dogs and buns to buy. Then you can figure out the number of packages of each.

The most obvious multiple of 10 and 8 is 80, but that's not the *least* common multiple. That number is 40: 10 × 4 = 40 and 8 × 5 = 40. Since hot dogs come in packages of 10, you'll need 4 packages of hot dogs. Since the buns come in packages of 8, you'll need 5 packages of hot dog buns.

Week Twenty-Eight

FRIDAY | PROBABILITY & STATISTICS

You have a spinner with equal-sized wedges. There are 3 green wedges, 3 red wedges, and 3 blue wedges. You spin the spinner. What is the probability that you will land on a blue wedge or a red wedge?

The fact that the wedges are equal in size matters. That means that there is an equal probability that you'll land on any of the wedges—making this problem relatively simple.

What you want to know is the number of opportunities to land on a blue or red wedge over the number of total opportunities to land on any color. How many blue or red wedges are there? That would be 6 total, right? And how many wedges are there in all? Nine.

So the probability of landing on a red or blue wedge is 6 out of 9 or $^6/_9$. This simplifies to $^2/_3$.

You can approach this problem in another way, too. There are 3 colors, with an equal number of wedges each. You want to know the probability of landing on either of 2 colors, which is 2 out of 3 or $^2/_3$.

The flexibility of math strikes again.

Week Twenty-Eight

SATURDAY | **LOGIC**

Your grandmother's kitchen is very strange. She has a 5-cup measuring cup and a 3-cup measuring cup, but no single- or 4-cup measuring cups. To make matters worse, all of the individual measurement markings have worn off the cups she does have. So you can't measure out 1 cup using just one of these measuring cups.

You are making bread and need to measure exactly 4 cups of flour. How can you do this with just 5-cup and 3-cup measuring cups?

First measure out 3 cups of flour and pour it into the 5-cup measuring cup. Then fill the 3-cup measuring cup with flour again. Pour as much as you can into the 5-cup measuring cup. This leaves 1 cup in the 3-cup measure. Now empty the 5-cup measuring cup. Pour the flour that is in the 3-cup measuring cup into the 5-cup measuring cup. (Remember? That's 1 cup.) Finally, fill the 3-cup measuring cup with flour and pour it into the 5-cup measuring cup. Ta-da! You've now got 4 cups of flour.

Of course, there is another solution. Fill the 5-cup measuring cup with flour. Pour as much as you can into the 3-cup measuring cup. That leaves 2 cups of flour in the 5-cup measuring cup. Now empty the 3-cup measuring cup. Pour what is in the 5-cup measuring cup into the 3-cup measuring cup. Fill the 5-cup measuring cup with flour. And finally, pour as much flour as you can from the 5-cup measuring cup into the 3-cup measuring cup. That leaves 4 cups in the 5-cup measuring cup.

A third solution is to get your grandmother a new set of measuring cups for her birthday.

Week Twenty-Eight

SUNDAY | GRAB BAG

Wyatt and Alex took turns driving on a trip to San Francisco. Wyatt drove the first 40 miles, and then Alex drove the rest of the way. On the way home, Wyatt drove first, and then Alex drove the last 50 miles. Who drove more and by how much? (Wyatt and Alex took the exact same route there and back.)

This problem seems impossible because you don't know how many miles they drove in all. But what *do* you know? First of all, you know that the trip to and from San Francisco is exactly the same. So they drove the same number of miles to the city as they did on the way home.

You also know that Wyatt drove 40 miles on the way there. Say Wyatt drove 10 miles on the way home. Since Alex drove 50 miles on the way home, the trip must be 60 miles long. That means Alex drove 20 miles on the way to San Francisco.

In this scenario, Wyatt has driven 40 + 10 or 50 miles, and Alex has driven 50 + 20 or 70 miles. Alex has driven 20 more miles than Wyatt.

But what if they need to drive farther to get to San Francisco? Assume that Wyatt drove 40 miles to San Francisco. What if he drove 30 miles to get home? Since Alex drove 50 miles to get home, the trip is now 80 miles long. And that means Alex drove 80 – 40 or 40 miles. This time, Wyatt drove 40 + 30 or 70 miles, and Alex drove 40 + 50 or 90 miles. Still, Alex has driven 20 more miles than Wyatt.

Try this out with other numbers. You'll always get the same result.

MONDAY | NUMBER SENSE

What is $\frac{1}{4} \div \frac{1}{2}$?

Who remembers how to divide fractions? Once you know the rule, the process is simple. Really! What you need to do is find the reciprocal of the second fraction. What's that? It's what you get when you turn a fraction upside down.

So turn $\frac{1}{2}$ upside down to get $\frac{2}{1}$. Now, you might automatically notice that this number is the same thing as 2. And it is. But leave it as a fraction. Now you can multiply: $\frac{1}{4} \times \frac{2}{1} = \frac{1 \times 2}{4 \times 1} = \frac{2}{4}$. (To multiply fractions, multiply the numerators and then multiply the denominators.)

Are you done? Yes and no. You can leave the fraction like this, but it's not in its simplest form. That's because the numerator and denominator are evenly divisible by the same number: $2 \div \frac{2}{4} \div 2 = \frac{1}{2}$.

So there's your answer.

Week Twenty-Nine

TUESDAY | ALGEBRA

What is another way to write the expression?

$$x(3x + 2)$$

The distributive property comes in play here. In other words, you'll distribute the single term to each of the terms in the binomial. So multiply x and $3x$, and then multiply x and 2: $3x^2 + 2x$.

WEDNESDAY | **GEOMETRY**

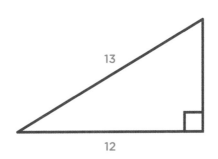

What is the length of the missing side?

This is a right triangle, so the Pythagorean theorem is the tool you'll use. First, you should identify the hypotenuse and the legs. The side that measures 13 is opposite the right angle, so that's the hypotenuse (or c).

The other two sides are the legs (a and b). It makes no difference which side you call a and which one you call b, so say that $b = 12$.

Substitute and solve for the variable: $a^2 + (12)^2 = (13)^2$ ➔ $a^2 + 144 = 169$ ➔ $a^2 = 25$ ➔ $a = 5$. Remember, the answer is positive because length is always positive.

Week Twenty-Nine

THURSDAY | APPLICATION

You are flying from San Francisco to Boston, where your aunt will pick you up at the airport. Your flight includes a layover in Chicago. The flight from San Francisco to Chicago is 4 hours long, and you leave at 8:45 a.m. The layover is 2 hours and 30 minutes, and the flight from Chicago to Boston is 1 hour and 45 minutes. Boston is 3 hours ahead of San Francisco. What time should your aunt be at the airport to pick you up?

Start by adding the first flight time to your departure time: 8:45 a.m. + 4 hours = 12:45 p.m. (Don't worry about the time difference; you can take care of that later.)

Add the layover time: 12:45 p.m. + 2 hours and 30 minutes = 3:15 p.m. Now add the flight time from Chicago to Boston: 3:15 p.m. + 1 hour and 45 minutes = 5:00 p.m. The last step is to add the 3-hour difference: 5:00 p.m. + 3 hours = 8:00 p.m. Of course your aunt may want to arrive a little later—to give you time to get off the plane and grab your bags.

Having trouble with the switch from a.m. to p.m.? Try using a 24-hour clock and then translate the time to a 12-hour clock at the end. (You'll get 20 hours at the end, which is of course, 8:00 p.m.) What about working in base 60? Picture a clock face, and count the fractions of the circles: 15 minutes is one-quarter of the circle, and so forth.

FRIDAY | PROBABILITY & STATISTICS

Your brother-in-law has bet you that the coin he has is fair. Out of 10 flips, the coin has come up heads 4 times. Should you take the bet?

Four out of 10 flips is pretty close to 50 percent. But you still shouldn't take the bet. The sample is way, way too small. Instead, ask him to repeat this experiment 20 times (which is 200 flips in all). Then pay close attention to what happens in each set of 10 flips.

For example, if the majority of each set of 10 flips comes up with fewer or more than 5 heads, it's likely that the coin is not fair. But if the results cluster around 5 heads, then you can assume that the coin is fair.

This is an example of the central limit theorem. The distribution of data will be normally distributed around the proportion parameter. (Yikes! Fancy math words.) In other words, the results you can count on will be clustered around a particular percentage or ratio. In this case, for the coin to be fair, the results should be clustered around 50% or $\frac{1}{2}$.

SATURDAY | LOGIC

A drunk man is walking home from a party. He has 60 yards to go before he gets to his house. Every minute, he lurches forward 3 yards but slips back 2 yards. How long does it take him to reach his house?

It might be tempting to say 60 minutes, because the man's net progress is 1 minute per yard. Unfortunately, that's not the correct answer. The goal is for the man to reach his house, which he will do at the 58th minute, just before he lurches back 2 yards.

Week Twenty-Nine

SUNDAY | GRAB BAG

Fill in each blank below with one mathematical symbol so that the equation is true. (No parentheses necessary.)

$$15 \underline{\quad} 10 \underline{\quad} 3 \underline{\quad} 7 = 43$$

So you've been given a big clue—that equal sign. Without those little horizontal bars, this problem is a lot harder. Each blank is a placeholder for a +, −, ×, or ÷.

Trial and error is probably going to be your best bet, but you can be smart about your choices. For example, multiplying each number will give you a number that's way, way too big. And all division would give you a number way too small. All subtraction gives a negative number. Even all addition is not going to get you to 43.

So you need to find a combination of operations. Start with addition—just for the heck of it: 15 + 10 = 25. If you multiply by 3, then you get 75. Subtract 7 and you get . . . not 43. It looks like division will need to be a part of this equation, but since you're looking for a whole number—43—the division must end up with a whole number.

Notice that 15 is evenly divisible by 3. You can't change the order of the numbers, but if you multiply 15 and 10, you get a multiple of 3. So, how about this: 15 × 10 ÷ 3 = 150 ÷ 3 = 50.

Now we're getting somewhere. The last operation is subtraction, because 50 − 7 = 43.

Week Thirty

How can you write 7 $\frac{3}{4}$ as a fraction?

Isn't this number already a fraction? Well, technically it's a mixed number. A fraction has no whole number, only a numerator and a denominator. As with many different math questions, you can approach this problem in several different ways.

A mixed number is basically a whole number plus a fraction. So 7 $\frac{3}{4}$ = 7 + $\frac{3}{4}$, which means you can change 7 to a fraction and then add. There are lots of fractions that equal 7, from $\frac{7}{1}$ to $\frac{56}{8}$. Which fraction should you use? Since you're adding to $\frac{3}{4}$, why not choose the fraction with 4 in the denominator? That means you'll add $\frac{28}{4}$ and $\frac{3}{4}$, which is $\frac{31}{4}$.

But there is another way to do this. The numerator of the improper fraction can be found by multiplying the denominator of the fraction by the whole number and then adding the numerator of the fraction: 4 × 7 + 3 = 31. The denominator of the improper fraction is the same as the denominator of the fraction in the mixed number.

Two ways, one solution.

TUESDAY | ALGEBRA

Combine like terms to simplify the expression. Then write the expression in descending order by degree. (The degree of a term is the exponent of that term.)

$$3x + 2x^2 + 8 + x^2 + 4x$$

What are the like terms in this expression? Think of the variables as objects. You have 3 x's, 2 x^2's, 8 no x's, 1 x^2, and 4 x's. Like terms have the same kinds of variables—or no variables at all. That means $3x$ and $4x$ are like terms. But $3x$ and $2x^2$ are not like terms.

You can add and subtract like terms. In this case, you'll get $7x + 3x^2 + 8$. But polynomials (expressions with more than one term) are typically written in descending order by degree. The degree of a term is the exponent. Therefore, $3x^2$ will go first, then $7x$, and lastly 8: $3x^2 + 7x + 8$.

Week Thirty

By what factor was rectangle *A* enlarged to create rectangle *B*?

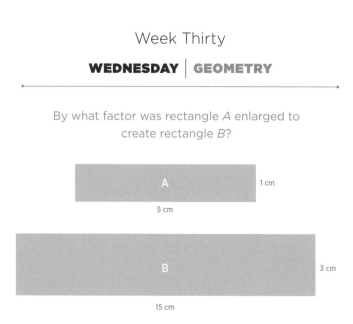

What you're looking for here is the scale factor. If a figure is shrunk, the scale factor will be less than 1. But if a figure is enlarged, the scale factor must be greater than 1. Since rectangle *A* was enlarged, the factor will be greater than 1.

Look at the dimensions given for each of the rectangles. In rectangle *A*, the dimensions are 1 centimeter and 5 centimeters. In rectangle *B*, the dimensions are 3 centimeters and 15 centimeters. In fact, the dimensions have changed by a factor of 3—in other words, each dimension was multiplied by 3 to get from rectangle *A* to rectangle *B*.

Xi is covering his kitchen floor with 1 foot by 1 foot tiles. If his kitchen looks like the diagram below, how many tiles will he need?

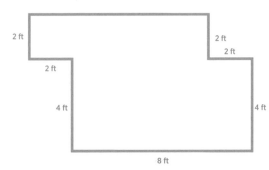

Want to draw it out? That's a great way to go. If you draw vertical lines to connect the top and bottom sides where the cutouts are, you'll get 3 rectangles.

The left rectangle is 2 feet by 2 feet, which means 4 tiles can fit inside. The middle rectangle—the largest—is 6 feet by 6 feet. Thirty-six tiles can fit inside that rectangle. The right rectangle is 2 feet by 4 feet. So 8 tiles will fit. Add up the number of tiles in each rectangle, and you'll get 4 + 36 + 8 = 48 tiles.

Week Thirty

This is a true story. On Halloween of 1936, *Literary Digest* predicted that Alfred Landon of Kansas would defeat President Franklin Delano Roosevelt in the November presidential election. The magazine had mailed out 10 million sample ballots to registered voters, or 25% of the electorate. Their mailing list was created with registers of telephone numbers, club membership rosters, and magazine subscription lists. Nearly 2.5 million people responded, and the magazine announced that Landon would win 57.1% of the popular vote and 370 Electoral College ballots. The next month, FDR won the election.

How did Literary Digest get its prediction so wrong?

Garbage in, garbage out, right? When the information is bad, the prediction will be bad, too, and that's what happened to *Literary Digest*. But how can you know that the information was bad? In 1936, the country was in a deep depression. Millions of people were out of work. The telephone was still a luxury that was out of reach for most Americans. Only the wealthy could afford club memberships and magazine subscriptions.

In short, the list was not representative of the entire population. The magazine had succumbed to selection bias. In order to accurately poll the American electorate, it would need to randomly select voters from a list of all voters, and then poll them.

But there was another problem as well. A mailed poll is not so simple to respond to. It's reasonable to assume that most folks would not take the time to respond and then mail back a sample ballot. This creates nonresponse bias.

So the structure of the poll is critical to its success.

SATURDAY | LOGIC

Six company presidents meet. They all shake hands with one another. How many handshakes are there in all?

If one of these folks is a germaphobe, the answer is easy: too many. But let's get down to the actual logic of this problem. Give each person a letter, so that you can distinguish between them all: A, B, C, D, E, and F. Of course, none of the people shake hands with themselves. (That would be weird.)

Person A shakes hands with everyone: AB, AC, AD, AE, AF. B already shook hands with A, leaving these combinations: BC, BD, BE, BF. C already shook hands with A and B, leaving these combinations: CD, CE, CF.

D already shook hands with A, B, and C, leaving these combinations: DE and DF. The only person thus far that E has not shaken hands with is F. And F has made all of the rounds. Count these pairs, and you get 15 handshakes.

You can also create a drawing to demonstrate the solution. Just be careful not to count a handshake more than once.

SUNDAY | **GRAB BAG**

Lagrange's four-square theorem goes like this: every positive integer can be written as the sum of 4 (or fewer) squares. For example, $8 = 2^2 + 2^2$.

What is the sum of squares for 54?

The easiest way to approach this problem is by trial and error. And to do that, you need to think of the numbers that add up to 54. It's a good idea to start with a perfect square. The largest perfect square that is less than 54 is 49 (or 7^2). To go from 49 to 54, you need to add 5, and the largest perfect square that is less than 5 is 4 (or 2^2). Now you've accounted for 53 out of 54. The only thing left is 1, which of course is 1^2. So the sum of squares for 54 is $49 + 4 + 1$ or $7^2 + 2^2 + 1^2$.

The last thing to check is whether or not you have the right number of squares. Lagrange says that you can have no more than 4, and you've got 3. Good to go.

Week Thirty-One

MONDAY | NUMBER SENSE

What is $5\frac{1}{3} + 3\frac{1}{4}$?

You have mixed numbers, which can be written as improper fractions (fractions with the larger number in the numerator). Or you can add the mixed numbers without changing them to improper fractions. And in this case, it makes perfect sense to do that. Why? Because when you add the fractional parts of the mixed numbers, you get a number that's less than 1. (Just think about measuring cups. If you pour $\frac{1}{3}$ cup of water in a 1-cup measuring cup, and then pour in $\frac{1}{4}$ cup of water, you won't overflow the cup.)

But before you add these numbers, you need to find a common denominator of the fractions. The least common multiple (LCM) of 3 and 4 is 12. So you can rewrite the numbers this way: $5\frac{4}{12} + 3\frac{3}{12}$. Add the whole numbers and then add the fractions to get $8\frac{7}{12}$. The fraction cannot be reduced, so you're all done.

Week Thirty-One

TUESDAY | ALGEBRA

Simplify the following expression:

$$6 + 3(x - 9)$$

When you simplify an expression, you want to have as few x terms and constants (the numbers by themselves) as possible. But in this case, you'll need to use your old friend the distributive property to distribute the 3 to the values inside the parentheses: $6 + 3x - 27$.

Now combine like terms—which in this case are only constants: $3x - 21$. If you want, you can factor 3 from both terms: $3(x - 7)$.

Week Thirty-One

WEDNESDAY | GEOMETRY

What is the area of the rectangle below?
What is the perimeter?

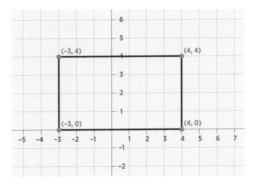

To find the area of a rectangle, you need to know its length and width. You can find this by simply counting the horizontal and vertical distances between each of the vertices above. The vertical distance is 4 units, and the horizontal distance is 7 units. Multiply these to find the area: 4 × 7 = 28 square units. Want another option? Count all of the boxes inside the rectangle.

The perimeter is the distance around the figure. Simply count around the rectangle, starting at one vertex to get 22 units. Or you can multiply the vertical distance by 2, multiply the horizontal distance by 2, and then add: 4(2) + 7(2) = 8 + 14 = 22 units.

Notice that the area of the rectangle is in square units, while the perimeter is in units. That's because perimeter is distance or length, which is never measured in square units.

A recipe calls for 4 $\frac{1}{2}$ cups of white flour and 2 $\frac{3}{4}$ cups of whole-wheat flour.

How much flour does the recipe call for?

Some folks can solve this problem in their heads. Others need a more concrete process. Start with the mental math. First, add the whole cups: 4 + 2 = 6 cups. Next, think about those partial cups. There are two $\frac{1}{4}$ cups in $\frac{1}{2}$ cup and three $\frac{1}{4}$ cups in $\frac{3}{4}$ cup—that makes five $\frac{1}{4}$ cups or $\frac{5}{4}$ cups or 1 $\frac{1}{4}$ cups. Add that to the 6 cups you already figured out to get 7 $\frac{1}{4}$ cups.

But if that process didn't work for you, try using the rules for adding mixed numbers. Start by getting a common denominator for the fractions. That is going to be 4, since both 2 and 4 are factors of 4. The first mixed number will change to $4\frac{2}{4}$. Now you can add: $4\frac{2}{4} + 2\frac{3}{4} = 6\frac{5}{4}$.

But that mixed number looks funny. It's not okay to leave an improper fraction as part of a mixed number. Think of $\frac{5}{4}$ as $\frac{4}{4} + \frac{1}{4}$ or $1 + \frac{1}{4}$. That means $6\frac{5}{4} = 7\frac{1}{4}$ cups.

As always, different processes can lead to the same answer.

Week Thirty-One

FRIDAY | PROBABILITY & STATISTICS

You have a spinner with 12 wedges. Half of the wedges are two times larger than the other wedges. You spin. What is the probability that you will land on a small wedge?

This probability problem is a little tough, because the wedges are not the same size. You can't simply say that the probability is 6 out of 12 or $\frac{1}{2}$.

The good news is that there are only two sizes. The better news is that the larger wedges are twice as large as the smaller wedges. That means you can divide the larger wedges in half to make 2 smaller wedges. With 6 larger wedges, you get 18 wedges in all.

But remember, there are only 6 of the original smaller wedges—or 6 opportunities to land on one. So the probability of landing on a smaller wedge is 6 out of 18. As a fraction, this is $\frac{6}{18}$ or $\frac{1}{3}$, a good deal smaller than the incorrect answer of $\frac{1}{2}$.

A square is divided into smaller squares. Then each of these smaller squares is numbered sequentially, left to right and top to bottom. If four of the smaller squares are numbered as shown below, how big is the square? Hint: these 4 squares are from the center of the larger square.

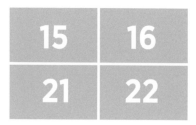

Since these squares are in the center of the larger square, there are the same number of rows above and below them, and the same number of columns to the left and right. The simplest approach may be to draw the larger square out from this center.

Add a row of 4 squares above and below, and a column of 4 squares to the left and the right. Fill in the numbers you can, and repeat the process of adding rows and columns until you have a completed large square. You'll find that the larger square is made up of 6 × 6 smaller squares.

Week Thirty-One

SUNDAY | GRAB BAG

Add a mathematical symbol between each of the numbers to make the statement true. (There is only one equal sign—the one given.)

$$1 \quad 2 \quad 3 \quad 4 \quad 5 \quad 6 \quad 7 \quad 8 \quad 9 \quad 10 = 1$$

Already, you know you're creating an equation. This is a big gift, since placing the equal sign would be much tougher. Because there is only one symbol between each number, you won't need exponents and roots. So you're left with the big four: +, –, ×, and ÷ . For the heck of it, why not start with addition and subtraction?

Clearly, if you add all of these numbers, you'll get an answer that is far bigger than 1. Therefore, there must be at least one subtraction symbol in there. Add the first 9 numbers: $1 + 2 + 3 + 4 + 5 + 6 + 7 + 8 + 9 = 45$.

That's way too big. (Subtracting 10 won't get you anywhere near 1.) What about adding the first 8 numbers? $1 + 2 + 3 + 4 + 5 + 6 + 7 + 8 = 36$. If you subtract 9, you get 27. Subtract 10, and you're at 17.

That's still way off, but look—8 + 8 is 16 and 16 + 1 is 17. If you subtracted 8 instead of adding it, you might very well be where you need to be.

Try it: $1 + 2 + 3 + 4 + 5 + 6 + 7 = 28$. Subtract 8, and you get 20. Subtract 9, and you're at 11. Subtract 10, and how about that? You're at 1.

So you need 6 addition signs, one between each of the first 7 numbers. The rest are subtraction.

MONDAY | NUMBER SENSE

Put the following numbers in ascending order:

$^2/_3$ 0.333 . . . $^4/_2$ $^3/_3$ 0.0125

I f these numbers were all decimals, you could compare the values in each place. If they were all fractions, you could find a common denominator and then compare the numerators.

Still, these are relatively common numbers, even if they do look funny in a list. Start by identifying the largest number. Both decimals have zeros in the ones place, so they are less than one. But one of the fractions is special, because the numerator is larger than the denominator. This means it's larger than 1. None of the other fractions have this quality, so $^4/_2$ is the largest number in the list.

Another of the fractions is special. Notice that in $^3/_3$ the numerator and denominator are the same. In fact, $^3/_3$ = 1. That makes it the second largest number.

The remaining numbers are less than 1. You may remember that 0.333 . . . is the same as $^1/_3$, which makes it less than $^2/_3$. And 0.0125 is certainly less than 0.333 . . . , because it has a 0 in the tenths place. That means that 0.0125 is the smallest, followed by 0.333 . . . , $^2/_3$, $^3/_3$, and $^4/_2$.

Week Thirty-Two

TUESDAY | ALGEBRA

Combine like terms to simplify the expression. Then write the expression in descending order by degree.

$$3x^2 + 9 - x + 12$$

What are the like terms in this expression? Turns out there are only two: 9 and 12. (Even though $3x^2$ and x both have x's, one is squared, so they're not like terms.) So it's pretty simple to simplify the expression: $3x^2 + 21 - x$. Now to reorder.

The first term has an x^2, so it's in the right place. Switch the 21 and x, being sure to keep the subtraction (or negative sign) with the x: $3x^2 - x + 21$.

WEDNESDAY | GEOMETRY

What is the area of the triangle? Can you find the perimeter? If so, how?

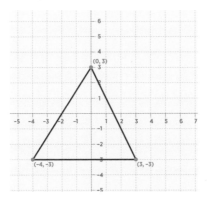

The formula for the area of a triangle is $\frac{1}{2}bh$, where b is the base of the triangle and h is the height.
The base is the length of the horizontal side or 7 units long. But what's the height? This is the vertical distance from the top vertex to the base. Since the top vertex is on the y-axis, you can simply count down to the base of the triangle. So the height is 6 units long. Plug everything into the formula, and you'll get 21 units squared.

You may have noticed that counting along the grid lines will only give you the lengths of horizontal or vertical sides. You can use the Pythagorean theorem to find the slanted sides of the triangle. The vertical line from the top vertex to the base of the triangle creates two right triangles. Find the lengths of the legs of these triangles and plug them into the Pythagorean theorem to find the hypotenuses (or is that hypotenii?). Add up the sides of the triangle to find the perimeter.

THURSDAY | APPLICATION

It's family reunion time, and you're in charge of the dinner rolls. You make great dinner rolls, but you need to quadruple your recipe, which calls for 5 $\frac{1}{2}$ cups of flour.

How many cups of flour do you need?

Here you go multiplying a mixed number by a whole number: 5 $\frac{1}{2}$ × 4. And as usual, there are different ways to approach it. The mental math for this problem is not too hard: 5 × 4 = 20 and $\frac{1}{2}$ × 4 = 2. Add to get 22 cups.

But if that process doesn't suit you, just change the mixed number into an improper fraction: 5 $\frac{1}{2}$ = $\frac{5 \times 2 + 1}{2}$ = $\frac{11}{2}$.

Then multiply by 4: $\frac{11}{2}$ × 4 = $\frac{44}{2}$ = 22 cups. Now get cookin'.

FRIDAY | PROBABILITY & STATISTICS

A nonprofit is hosting a raffle. Each ticket has a number printed on it, ranging from 535201 to 535500. If all of the tickets are sold, and you purchase 10 tickets, what is the probability that you will win the raffle?

How many tickets should you have purchased to have a $^{50}/_{50}$ chance of winning?

Like all probability problems, you want to know the ratio of the total positive outcomes to the total outcomes. You have purchased 10 tickets, so the total positive outcomes is 10. (In other words, 10 of the total tickets are yours.) But how many tickets are there in all? The tickets are numbered from 535201 to 535500. The first three digits of the ticket numbers are the same, so the last three digits will tell you how many tickets were sold. Since the range is 201 to 500 inclusive, there were 300 tickets sold.

That means that 10 of the 300 tickets were yours, and the probability of your winning is $^{10}/_{300}$ or $^{1}/_{30}$. A $^{50}/_{50}$ chance of winning is the same as $^{1}/_{2}$, so you would have needed to purchase 150 tickets to get that really great chance of winning.

Week Thirty-Two

SATURDAY | LOGIC

Stella works on a farm in the summer. The farmer wants her to build 4 pens for his 9 chickens. He also wants there to be an odd number of chickens in each pen.

How can Stella do this?

You probably noticed that 4 separate pens won't hold an odd number of chickens each. If you put 3 chickens in each pen, you'll have no chickens for the fourth pen. And if you put 1 chicken in each of 3 pens, you'll have 6 chickens in the fourth pen. None of these combinations works.

But you can build 3 pens, and put 3 chickens in each. Then build the fourth pen around the 3 smaller pens. That fourth pen has 9 chickens, which is also an odd number.

Week Thirty-Two

SUNDAY | GRAB BAG

You have more than $1 in change, but you can't make change for $1. What is the most amount of money you could have? (U.S. denominations, please.)

No U.S. coin is greater than $1, so the one-coin answer is out. Also notice that there is no restriction on the number of coins you must have. It might help to think about typical ways you can make change for $1. The most obvious may be 4 quarters. So, the answer cannot have more than 3 quarters.

In fact, start with those 3 quarters. That gives you 75¢, which is less than $1. Add on, starting with dimes: 0.75 + 0.10 = 0.85, 0.85 + 0.10 = 0.95, and 0.95 + 0.10 = 1.05. Now you're above $1. You can't make change for $1, but is this the largest amount of money you could have?

What would happen if you added nickels? You'd be able to make change for $1, that's what. Two dimes plus 1 nickel is 25¢, and 75¢ + 25¢ is $1. So no nickels.

But could you add another dime? Sure thing. You only have 3 dimes. Adding 1 more will give you $1.15, and you can't make change for a dollar with 3 quarters and 4 dimes.

You can add pennies, though, but only 4 of them. If you add 5 pennies, you can make change for a dollar. At this point, you're up to 3 quarters, 4 dimes, and 4 pennies. That's a total of $1.19.

Week Thirty-Three

MONDAY | NUMBER SENSE

This list is part of a sequence. What is the next number?

$$1\tfrac{1}{3},\ 2,\ 2\tfrac{2}{3},\ 3\tfrac{1}{3},\ 4,\ 4\tfrac{2}{3},\ \ldots$$

If you think of this sequence methodically, you can find the pattern and the next 50 numbers (if you wanted to go that far). Since this is only part of the sequence, you know that the first number is not $1\tfrac{1}{3}$. Still, it makes sense to look at the change from one number to the next.

Start by looking at 2 and $2\tfrac{2}{3}$. That's pretty straight-forward: $\tfrac{2}{3}$ was added. But is this the pattern? Instead of looking at the next two numbers, take a look at 4. To get to $4\tfrac{2}{3}$, add $\tfrac{2}{3}$. So far, so good. Now you've got to consider the more complicated pairs, to make sure that the pattern holds.

For example, to get from $1\tfrac{1}{3}$ to 2, you are also adding $\tfrac{2}{3}$. You don't need to do any fancy arithmetic here. Just think about the numbers. Add $\tfrac{1}{3}$ to get to $1\tfrac{2}{3}$; add another $\tfrac{1}{3}$ to get to 2.

So, the pattern is really simple, even if the numbers look ugly—just add $\tfrac{2}{3}$. But how do you find the next number in the sequence? You're going to add $\tfrac{2}{3}$ to $4\tfrac{2}{3}$. If you add $\tfrac{1}{3}$ you'll get to 5. Add another $\tfrac{1}{3}$, and you'll be at $5\tfrac{1}{3}$, which is the next number in the sequence.

For $-5x - 27 = 9 + x$, can $x = -3$?

There are a couple of ways to find out. You can solve for x and see if it's -3. To do this, you need to combine like terms, which means getting all of the x's on one side of the equation and all of the constants (numbers that are all by their lonesome) on the other side of the equation. First, deal with the variables: $-5x - x - 27 = 9 + x - x$ ➜ $-6x - 27 = 9$. Notice that you end up adding two negatives, and that's where careless mistakes can happen.

Next, add 27 to both sides of the equation: $-6x - 27 + 27 = 9 + 27$ ➜ $-6x = 36$. Finally, divide each side of the equation by -6: $x = -6$. So -6 is the solution to this equation, not -3.

You can also substitute and see what happens: $-5(-3) - 27 = 9 + -3$ ➜ $15 - 27 = 6$ ➜ $-12 = 6$. Since you don't get the same thing on both sides of the equation, $x \neq -3$.

Week Thirty-Three

WEDNESDAY | GEOMETRY

What are the area and the perimeter?

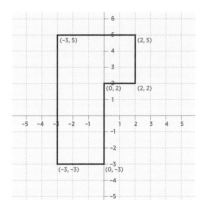

It might be easiest to start with the perimeter. Just find the length of each side (by counting the units between the vertices) and then add those lengths together, starting at the top and going clockwise: 5 + 3 + 2 + 5 + 3 + 8 = 26 units.

The area takes a little more thinking. You'll need to divide the figure up into smaller rectangles, find the area of each of those rectangles, and then add. And of course there are many different ways to divide this figure into rectangles.

How about drawing a horizontal line from (0, 2) to the longest side of the figure? (Go ahead and draw on the picture if it helps.) This gives you two rectangles. The one on top has dimensions of 5 and 3. The one on the bottom has dimensions of 5 and 3. (Coincidence!) So the area of each of these rectangles is 15 square units. The total area is 30 square units—or the sum of the areas of the smaller rectangles.

Week Thirty-Three

THURSDAY | APPLICATION

You have a large, cube-shaped planter for your front porch. Because it's a cube, the sides all have the same measure: 15 inches. How many square feet of potting soil will you need to fill it? (Assume that the potting soil will go all the way to the top. The volume of a cube is *lwh*.)

There are a few starting points you can take with this problem. You can find the volume in cubic inches first. Or you can convert 12 inches to feet and then find the volume in cubic feet. Since the potting soil is measured in cubic feet, that's the way to go.

The pot is 15 inches by 15 inches by 15 inches— or 1.25 feet by 1.25 feet by 1.25 feet. So the volume of the cube is 1.95 cubic inches. You'll need just about 2 cubic feet of potting soil.

But what if you take the other option: finding the volume of the cube in cubic inches first? That would be 3,375 cubic inches. To convert to cubic feet, you need to divide by 12^3 or the number of cubic inches in 1 cubic foot. (You might think it's enough to divide by 12, but remember, you're dealing with the volume of a three-dimensional figure.)

Since 12^3 is equal to 1,728, that's what you'll divide by: 3,375 ÷ 1,728 = 1.95. Same answer as above.

Week Thirty-Three

There are 12 socks in your sock drawer. Eight are white, 3 are black, and 1 is green. (Not all of the socks have mates.) You pull out 1 white sock, and then you fish around for another sock. What is the probability that you choose a white sock the second time?

Probability can be difficult to get your head around. But if you think of the problem systematically, you can figure it out no problem.

When you choose the first sock, how many are left in the drawer? There were 12, and now there are 11. And when you choose the first white sock, how many white socks are left in the drawer? There were 8, so now there are 7.

The probability that you will choose a white sock is 7 white socks out of 11 total socks or $^7/_{11}$. That wasn't so bad, was it?

Week Thirty-Three

SATURDAY | LOGIC

Not all sweets are healthy. A cookie is a sweet.
Which of these conclusions is true?

1. A cookie could be healthy.
2. All sweets are not healthy.

It turns out that only the first conclusion is true. Since not all sweets are healthy and a cookie is a sweet, a cookie may or may not be healthy. But because not all sweets are healthy, some sweets may be healthy. This means that not all sweets are not healthy.

SUNDAY | **GRAB BAG**

What is the next number in this sequence?

4, 6, 12, 18, 30, 42, 60, 72, 102, 108, . . .

Don't spend too much time looking for an algebraic pattern. The answer lies in the types of numbers that are missing. All of the numbers in this list are *composite numbers*, which means they are not prime numbers. However, each of these numbers is exactly between two prime numbers. For example, 4 is between 3 and 5— both prime numbers. And 18 is between 17 and 19— both prime numbers.

Now that you understand the pattern, you can find the next number. It might be helpful to grab a list of prime numbers. The largest prime number associated with this sequence is 109. (That's because 108 lies between 107 and 109.) So you're looking for two prime numbers that are greater than 109 and are two units apart.

Those numbers are 137 and 139, which means the correct answer is 138. For another challenge, try finding the next number in the sequence.

Week Thirty-Four

MONDAY | **NUMBER SENSE**

What are the next two numbers in this sequence?

0, 5, 3, 8, 6, 11, 9, . . .

This sequence is a bit unusual, because the same rule doesn't apply to each pair of numbers. You can see that because the numbers increase and then decrease in pairs. So think about the increasing pairs. What is one way to describe going from 0 to 5? Add 5, right? And how do you get from 3 to 8? Add 5. (Seeing a pattern here?) What about 6 to 11? Add 5 again.

Now, take a look at the pairs of decreasing numbers. How do you get from 5 to 3? Subtract 2. And 8 to 6? Subtract 2. Finally, 11 to 9? Subtract 2. So there is a pattern; it's just not consistent between each pair of numbers. First add 5 and then subtract 2. That means that the next two numbers will be 9 + 5 = 14 and 14 – 2 = 12.

There is another way to look at this sequence though. Take a look at the pairs created by every other number:

0 and 3 → 0 + 3 = 3;

5 and 8 → 5 + 3 = 8;

3 and 6 → 3 + 3 = 6;

8 and 11 → 8 + 3 = 11;

6 and 9 → 6 + 3 = 9;

11 and ? → 11 + 3 = 14; and

9 and ? → 9 + 3 = 12.

Love that flexible thinking.

Week Thirty-Four

TUESDAY | ALGEBRA

What is the solution set for $2x - 6 < 2$?

Wait. What's a *solution set*? It's a set of numbers that makes a problem true. Equations generally have one, two, or no solutions. But this is an inequality (like this less-than problem), so there is a range of solutions.

The great news is that you can start by isolating x, just like you would do with an equation. Add 6 to both sides of the inequality to undo the subtraction of 6 on the left side: $2x < 8$. Just keep that less-than sign right where it is. Now divide each side by 2: $x < 4$. And that's the answer.

But what does it mean? Every single number in the world that is less than 4 is a solution to this problem. That includes fractions and decimals and negative numbers. Try and see. If $x = -1$, does the inequality work? $2(-1) - 6 < 2 \rightarrow -3 - 6 < 2 \rightarrow -9 < 2$. Yep, -9 is less than 2, so $x = -1$ is indeed in the solution set.

What are the area and circumference of the circle below?

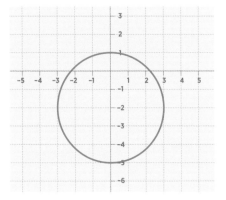

Both of the formulas require the radius of the circle. To find the radius, you need to identify the center of the circle. Notice that at $y = -2$, the x-values of points on the circle are both -3.

This means that the y-axis cuts the circle into two equal parts. So the x-value for the center will be 0. The y-value will be exactly between 1 and -5. That's -2. So the center is at $(0, -2)$.

At $(0, -2)$, there are exactly 3 units to any point on the circle. (It helps to look at the horizontal and vertical distances to verify this.) This means the radius is 3.

Now that you have the radius, it's easy to plug it into the formulas. $A = 3.14 \times 3^2 = 3.14 \times 9 = 28.26$ square units. $C = 2 \times 3.14 \times 3 = 18.84$ units.

Week Thirty-Four

THURSDAY | APPLICATION

Dan's net income is $47,530 per year. He wants to purchase a house, spending 25% to 30% of his income on a mortgage.

How much should his monthly mortgage be?

The answer is going to be a range, since Dan's percentages are a range.

First find 25% of $47,530: 0.25 × 47,530 = $11,882.50. Then find 30% of $47,530 = 0.3 × 47,530 = $14,259.

But these aren't the monthly payments he's shooting for. Thank goodness! Divide each of these by 12 to get the monthly range:

11,882.50 ÷ 12 = $990.21 and 14,259 ÷ 12 = $1,188.25.

That's better. Between $990.21 and $1,188.25 is much more comfortable for a monthly mortgage payment.

Week Thirty-Four

FRIDAY | PROBABILITY & STATISTICS

Lily averaged all of her students' 5 test scores, and then she realized that she misplaced Kyla's last test. Kyla's test score average is 86.2. Her other test scores were 75, 87, 99, and 90. Each score is equally weighted.

What is Kyla's missing test score?

To find the average of the test scores, Lily added the scores and divided by 5. (Why 5? Because there are 5 tests, and each of the tests is equally weighted.)

She knows that Kyla's average is 86.2. So she can work backwards to find the missing test score. Since she divides the sum of the scores by 5, she can multiply the average by 5 first: $86.2 \times 5 = 431$. Now she can add up all of the test scores that she has: $75 + 87 + 99 + 90 = 351$. Finally, she can subtract 351 from 431: $431 - 351 = 80$. The missing test score is 80.

Is all of this confusing? Another way to look at the problem is to set up an equation, with x as the missing test score: $(x + 75 + 87 + 99 + 90) \div 5 = 86.2$. Then solve for x.

Georgia has forgotten the combination to her safe. There are three unique numbers, and she remembers these four clues: The second number is four times the first number and twice the third number. The third number is twice the first number. The third number is a perfect square. The first number has one digit, and the other two numbers have two digits.

The answer is 8, 32, 16, but how do you get there? The biggest clue is the last one, that the first number has one digit, but the others have two. There are so few one-digit numbers that you can test each one.

Or you can think about the second and third clues. The third number is both twice the first number and a perfect square. Make a list of the perfect squares: 4, 9, 16, 25, 36, . . . Only four of these are possibilities, since the perfect square must be double another number: 4, 16, and 36. The third number cannot be 4, since it has to have two digits. Therefore, there are only two choices: 16 and 36.

Now consider the first clue, using these two choices. If the last number is 16, the second number is 32. (The second number is two times the third number.) That makes the first number 8. (The first number is four times the second number.) These three numbers fit all of the criteria, but to be sure, test 36 as well.

If the last number is 36, the second number is 72. And if the second number is 72, the first number must be 18. Ah, but the first number should only have one digit, so the previous solution (8, 32, 16) is correct.

Week Thirty-Four

SUNDAY | GRAB BAG

Place an addition, subtraction, multiplication, or division symbol between each of the following numbers to get 40:

$$4 \quad 6 \quad 8 \quad 8$$

Note: You can use any of the operations more than once, but you can only use the numbers one time each. You can also rearrange the numbers.

This is one of those times when a math problem doesn't have only one right answer. So yours might differ from the one given here. As long as you can add, subtract, multiply, and/or divide the numbers above and get 40, you're good to go. One answer is 4 x 6 + 8 + 8 = 40.

Here's another solution. This time you'll need to rearrange the numbers. Multiply 8 by 8 to get 64. Then multiply 4 and 6 to get 24. Finally, subtract 24 from 64 to get 40: $(8 \times 8) - (4 \times 6) = 40$.

MONDAY | NUMBER SENSE

Write 4.3×10^{-3} as a number.

If you've ever worked with really small or really large numbers, or if you remember way back to your math and science classes, you will recognize this as *scientific notation*. But how do you express scientific notation as a number? The biggest clue is that little tiny number beside the 10. This exponent tells you what to do with the decimal point in the two-digit number. Because the exponent is negative, you'll move the decimal point to the left, making a smaller number. And you'll count 3 places, because the exponent is -3. This means that 0.0043 is the same thing as 4.3×10^{-3}.

What if you can't remember whether to move the decimal point to the right or to the left? Think about what happens when you increase a positive exponent. The value gets larger and larger, right? In other words, 10^4 is larger than 10^3. You can then reason that when the exponent gets smaller, the number gets smaller. So 10^2 is smaller than 10^3. Negative exponents must create really small numbers. And in fact, when a number is raised to a negative exponent, you'll get a number that's less than 1, which is exactly what happens when the decimal point is moved to the left of the decimal in scientific notation.

Week Thirty-Five

TUESDAY | ALGEBRA

What is the solution set for $8x - 5 > 5x - 2$?

I n this problem, you've got variables on both sides of a greater-than symbol. Remember, treat the problem as if it's an equation, and you'll be just fine. Start by getting $8x$ by itself on the left side of the inequality. Do that by adding 5 to each side, which undoes the subtraction: $8x > 5x + 3$. (You added 5 to -2, which gives you +3.) Next, get the 3 by itself on the right side of the inequality, by subtracting $5x$ from both sides: $3x > 3$. Now you can divide each side of the inequality by 3 to find that $x > 1$.

So what's the solution set? Any number that is bigger than 1, including fractions and decimals.

Week Thirty-Five

WEDNESDAY | GEOMETRY

A circle is graphed on a coordinate plane. It has a radius of 4 units and goes through the points (-1, 6) and (-1, -2). What are the coordinates of the center of the circle?

If you're a visual person, it might be helpful to draw a picture of this circle. Notice that the two given points have the same *x*-values. This means that when the points are connected, this line segment is vertical. In fact, the length of this line segment is the diameter of the circle. So the center of the circle is at the midpoint of this line segment.

Since the center lies on this vertical line segment, its *x*-value is -1. To find the *y*-value, you only need to find the difference of the *y*-values and divide by 2: 6 – -2 = 8 and 8 ÷ 2 = 4. The *y*-value is 4. So the point (-1, 2) is the center of the circle.

THURSDAY | **APPLICATION**

A pet shelter spays and neuters cats and dogs, plus gives them a checkup. The shelter wants to spend no more than $2,000 per day on these services. The base cost for vet services for any number of animals is $250 per day. Each spay or neuter costs $75. How many pets can the shelter spay, neuter, and give a checkup to each day?

One way to approach this problem is to set up an equation and solve it. The variable is the number of pets, which is multiplied by $75 to find the total cost of spaying and neutering. Then add $250 for the daily vet fee, and set the whole thing less than or equal to $2,000, the amount the shelter wants to spend each day: $75x + 250 \leq 2,000$.

(Why use a less-than-or-equal-to symbol? Because you want the daily cost to be no more than $2,000.)

Now solve for x. Start by subtracting 250 from both sides of the inequality: $75x + 250 - 250 \leq 2,000 - 250$ → $75x \leq 1,750$. Now divide each side of the equation by 75: $75x \div 75 \leq 1,750 \div 75$ → $x \leq 23.33$. There's no such thing as 0.33 of a cat or dog, so the shelter can spay and neuter no more than 23 animals while staying under budget.

Of course, you can also solve this problem mentally, starting by subtracting $250 from $2,000. Then divide by $75 to find out that, at about 23 cats and dogs, the shelter would reach its $2,000 limit.

Week Thirty-Five

FRIDAY | PROBABILITY & STATISTICS

A number is randomly selected from 1 to 10. What is the probability that the number is odd?

Probability is all about outcomes. In this case, the probability is the ratio of the number of odd numbers between 1 and 10 to all of the numbers from 1 to 10. This ratio is often described as a fraction.

So in this case, you need to know how many odd numbers there are from 1 to 10. Turns out there are 5 odd numbers: 1, 3, 5, 7, and 9. And there are 10 numbers in all: 1, 2, 3, 4, 5, 6, 7, 8, 9, and 10. So the probability that an odd number is selected is 5:10 or $\frac{5}{10}$, which of course can be simplified to $\frac{1}{2}$.

And here's a bonus question: What is the probability that the number selected will be even?

Week Thirty-Five

SATURDAY | LOGIC

After her death, your great-grandmother left you a priceless diamond ring. You want to have the ring appraised, and you've found the best gem appraiser in the country. Problem is she's on the other side of the country. To ship the ring, you've purchased a box that can be fitted with multiple locks. You have several locks (and their keys), but the appraiser does not have any of the keys to those locks. If you send a key to the appraiser, it could be duplicated. How can you send the ring to the appraiser securely, and how can the appraiser send the ring back to you securely?

What a wonderful problem to have, right? Typically, a ring would be sent to the appraiser and then back to you, but that won't work in this situation. What if you sent the box back and forth multiple times?

Check this out: Send the box with one of your locks on it. When the appraiser gets the box, she attaches her own lock and sends it back to you. Then you can remove *your* lock, and send it back to the appraiser, who then removes *her* lock to access the diamond. Repeat the process (in reverse) to get the diamond back to you after the appraiser is done.

Week Thirty-Five

SUNDAY | GRAB BAG

Four consecutive whole numbers have a sum of 50. What are those numbers?

Consecutive whole numbers are one apart from one another. For example, 4 and 5 are consecutive, as are 109 and 110. In this case, you want to find 4 consecutive whole numbers that have a sum of 50.

A good place to start is to divide 50 by 4. Why? Since these 4 numbers are consecutive, they will likely be close to 50 ÷ 4 or 12.5. Now you can guess and check: 12 + 13 + 14 + 15 = 54. That's too large, so how about trying 11 + 12 + 13 + 14? That's exactly 50, and you've found the answer.

MONDAY | NUMBER SENSE

What is the next number in this list?

0, -3, 6, 33, 96, . . .

Here again you need to find the pattern by figuring out how you get from one number to the next. There are multiple ways to get from 0 to -3, but which one also gets you from -3 to 6 and from 6 to 33? In this case, it might be helpful to look at the second and third pairs of numbers in the list.

Going from -3 to 6, you can add 9. But that won't get you from 6 to 33. Some bigger operation is happening, since the numbers are growing quickly. What about an exponent? If you square -3, you get 9. To get to 6, subtract 3.

Does that rule apply to the other pairs? Check 6 and 33. Square 6 to get 36. Then subtract 3 to get 33. Yep! You've found the pattern. (If you want to be absolutely sure, check all of the other pairs.)

So to find the next number in the list, square 96 to get 9,216. Then subtract 3 to get 9,213, which is the last number in the list.

Which of the following is a graph of $x \geq 3$?

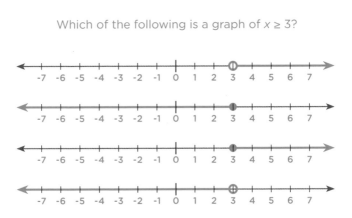

There's no other way around it; you've got to understand what that little symbol means: $\geq$. Think of it this way: it's a mix of an equal sign and a greater-than sign. So x is greater than or equal to 3. With this information, you can start eliminating the graphs. Because it's *greater than* or equal to, the arrow is going to go to the right, indicating that the numbers that are greater than 3 are included. And because it's greater than or *equal to*, the dot at 3 has to be filled in. An open dot means the number is not included. So the third graph from the top is the correct one.

WEDNESDAY | GEOMETRY

A circle is graphed on a coordinate plane. It has a center at (-4, -4) and a point at (2, -4). What is the diameter of the circle?

Drawing what you know may help you get your bearings, but you really only need to know the coordinates of the points given. These two points fall on a horizontal line, because the *y*-values are the same. Since one of the points is the center and the other is a point on the circle, the horizontal distance between them is the radius. To find the distance between these points, find the difference between the *x*-values: 2 – -4 = 6. The radius is 6, so 12 units is the diameter of the circle.

THURSDAY | APPLICATION

A recipe calls for 1 ¼ cups sugar. You only have ¾ cup of sugar. By what factor will you need to change the amounts of the other ingredients in the recipe?

Do you have ½ or ¾ the amount of sugar you need? Who knows, but that's the fraction you're looking for. Here is one way to think of this question: ¾ is how much of 1 ¼? That little word "of" means multiply. Want another clue? That little word "is" means equals. So you can create an equation pretty easily: $3/4 = $ what $ \times 1 1/4$.

If those fractions are freaking you out, don't worry. It might help to write the mixed number as an improper fraction. It also may help to use a variable: $3/4 = 5/4x$. Now you just need to undo the multiplication to get the variable by itself. This means multiplying each side of the equation by $4/5$: $3/4 \times 4/5 = 5/4x \times 4/5$ ➜ $12/20 = x$. This is not a useful fraction, so simplify: $12/20 = 3/5$.

Whew! So this means you'll be making ⅗ of the original recipe.

FRIDAY | **PROBABILITY & STATISTICS**

It's time for the annual Tinkertown Turtle races! The day culminates in the 10-meter race, with gold, silver, and bronze winners. This year there are five entries. In how many different ways can the turtles place?

You're being asked to find the number of ordered combinations. The order matters because the race has a first, second, and third place. There are 5 opportunities for a first-place winner. But after the first-place winner is decided, the second-place winner will come from the remaining 4 entries, and then the third-place winner will come from the remaining three entries. So the total number of ways that the turtles can place in the 10-meter race is 5 x 4 x 3 or 60.

A specialty store sells a calendar made up of two cubes. These cubes can be arranged to show the current day of the month. What single-digit numbers are on the cubes?

There are only 31 combinations needed, and there are 12 faces between both cubes. The numbers 0, 1, 2, and 3 must be on one cube, so that the first digits can be represented. The numbers 0, 1, 2, 3, 4, 5, 6, 7, 8, and 9 are needed for the second digits, but the other cube only has six sides. Plus, there are two dates with identical digits: 11 and 22.

Since 0, 1, 2, and 3 are on one cube, you need 1 and 2 on the other cube too. Put 4 and 5 on the first cube, which means you need a 0 on the second cube (to make 04 and 05). Now you've got six numbers taken care of on the first cube, and three numbers on the second cube. What's left? The remaining second digits of the dates: 6, 7, 8, and 9. But you only have three slots. Time to get creative. If drawn a certain way, a 6 can be turned upside down to make a 9. So the second cube will have 0, 1, 2, 6, 7, and 8.

To summarize: In this solution, the first cube has 0, 1, 2, 3, 4, and 5, while the second cube has 0, 1, 2, 6, 7, and 8. That's 12 faces, six numbers per cube. Sure, you can make nonexistent dates like 00 and 32, but all of the real dates are represented.

There is another solution to this particular problem. See if you can find it.

Week Thirty-Six

SUNDAY | GRAB BAG

Insert two mathematical symbols to make this problem true:

$$16 - 4 = 2$$

There are already two mathematical symbols in this problem, a minus sign and an equal sign. And each of the numbers is separated by a mathematical symbol, so it doesn't make sense to insert one of the four common mathematical symbols: +, −, ×, or ÷.

What symbols are left? You could raise these numbers to a power, like 2 or 3. Squaring 16 and 4 makes the value on the left too large. Squaring 4 and 2 makes the value on the right too large. But what about the inverse of raising a number to a power, taking the root? The most common root is the square root. Why not try there? Luckily both 16 and 4 are perfect squares. The square root of 16 is 4 and the square root of 4 is 2. Subtract these, and voilà, you get 2. So the two mathematical symbols you'll add are two square-root signs, one over the 16 and one over the 4.

Week Thirty-Seven

MONDAY | NUMBER SENSE

Write 8.63×10^5 as a number.

Again, you're rewriting scientific notation as a number. This time, the exponent is positive, so the decimal point will be moved to the right. How many places will it be moved? Since the exponent is 5, you'll move the decimal point 5 places to the right. Two of these are already written in 8.63. The other three places need to be added—with zeros. So 863,000 is the same as 8.63×10^5.

TUESDAY | ALGEBRA

Solve for x.

$$7x - 3 = 3 + 6x$$

Solve for x, and you're done. This time, why not start by isolating the 3 on the right side of the equation? Do this by subtracting $6x$ from both sides: $x - 3 = 3$. Now you might be tempted to subtract 3 from both sides of the equation. Don't do it! You need to get rid of the 3 on the left side, which means *adding* a 3, to undo the subtraction: $x = 6$. No division is necessary this time, because x is already isolated.

The two triangles below are congruent. What does that say about their corresponding parts? List any congruence statements you can make.

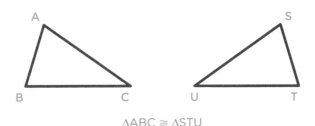

$\triangle ABC \cong \triangle STU$

N otice that the triangles are not oriented in the same way. But if you pay close attention to where the angles and sides are, you'll find that you can write six congruence statements—three with angles and three with sides. The congruent angles statements are $\angle A \cong \angle S$, $\angle B \cong \angle T$, and $\angle C \cong \angle U$. The congruent sides statements are $\overline{AB} \cong \overline{ST}$, $\overline{BC} \cong \overline{TU}$, and $\overline{AC} \cong \overline{SU}$.

Week Thirty-Seven

THURSDAY | APPLICATION

For your Halloween costume, you purchased 6 ½ yards of fun fur fabric. (It was on sale!) You used ⅔ of the fabric. Now you have an idea to make a hat and muff for your sister. You need 2 ⅓ yards of fabric. Do you have enough?

This problem requires multiplying fractions. And there are some mixed numbers thrown in there. Hoo-boy! First, you need to know how much fabric you have left from the original amount. You used ⅔ of the 6 ½ yards, so you could multiply these and then subtract the answer from 6 ½. Or you could just multiply 6 ½ by ⅓, which is the fraction of fabric left over. But how do you multiply a mixed number by a fraction? One approach is to change the mixed number into an improper fraction: $6 \frac{1}{2} = \frac{6 \times 2 + 1}{2} = \frac{13}{2}$. Now all you need to do is multiply: $\frac{13}{2} \times \frac{1}{3} = \frac{13}{6}$.

And then it might be helpful to change this improper fraction into a mixed number, so that you can compare to 2 ⅓. To do this, divide 13 by 6 to get 2. Then put the remainder in the numerator of a fraction and keep the denominator. So you have 2 ⅙ yards of fabric left over. Not quite enough for the hat and muff. Boo!

Week Thirty-Seven

FRIDAY | PROBABILITY & STATISTICS

You flip a coin 3 times. What is the probability of getting tails all 3 times?

This is an *and* question. You want to get tails *and* tails *and* tails. And this is where probability veers from what you may have learned about word problems. Instead of adding the probabilities here, you will need to *multiply*.

The probability of getting tails is $\frac{1}{2}$. That's because there are 2 outcomes in all (heads or tails) and 1 favorable outcome (tails). It doesn't matter how many times you flip that coin, the probability of getting tails in that one flip is always $\frac{1}{2}$. So the probability of getting 3 tails in a row is $\frac{1}{2} \times \frac{1}{2} \times \frac{1}{2} = \frac{1}{8}$.

Week Thirty-Seven

SATURDAY | LOGIC

Three gangsters are arrested and questioned by the police. They make the following statements:

Mr. Pink: Mr. White is a rat.
Mr. White: Mr. Brown is a rat.
Mr. Brown: Mr. White is a liar.

The police officers know that rats lie, but all other gangsters tell the truth. They also know there is only one rat. Who is the rat?

Did you notice that both Mr. Pink and Mr. Brown had something to say about Mr. White? If Mr. Pink is right, then Mr. White is the rat (and he's lying). If Mr. Brown is right, then Mr. White is lying, so he must be the rat.

So Mr. White is the rat, and Mr. Pink and Mr. Brown are telling the truth.

Week Thirty-Seven

SUNDAY | **GRAB BAG**

Find three consecutive whole numbers with this property: their sum is equal to their product.

The sum is what you get when you add numbers together. The product is the result of multiplication. So you want three consecutive whole numbers that add and multiply together to get the same number. In other words: $x + y + z = xyz$, where x, y, and z are consecutive whole numbers.

Trial and error might be the best approach here—as long as you're making smart choices. Choosing really large numbers doesn't make sense. The larger the numbers you are multiplying, the larger the product—and the further the product is from the sum. So what about one-digit numbers, like 4, 5, and 6? Their product is 120 and their sum is 15. The product is way larger than the sum, so you should go even smaller. What about 3, 4, and 5? That gives you a product of 60 and a sum of 12. Still a huge difference between the two. Time to go even smaller: 1, 2, and 3. Their product is 6 and their sum is 6.

Week Thirty-Eight

MONDAY | **NUMBER SENSE**

What is the tenth value in the list of numbers below?

$$1, 3, 3, 9, 27, \ldots$$

To find the tenth value of the sequence in question, you need to find the *rule*, or what happens to generate each of the numbers. So take a closer look. To get from 3 to 3, you multiply: $1 \times 3 = 3$. Does this rule hold up? Try it out! $3 \times 3 = 9$, $3 \times 9 = 27$. So here's the rule: multiply the previous two numbers to get the next number.

With this you can go ahead and generate the sequence to the tenth value. You've got five numbers already. Now you need to find the next five: $9 \times 27 = 243$; $27 \times 243 = 6,561$; $243 \times 6,561 = 1,594,323$; $6,561 \times 1,594,323 = 10,460,353,203$. And that last number? It's the tenth value—and HUGE!

Solve for x:

$$3(2x + 4) = 4x + 10$$

There's a lot going on here. You've got x's on both sides of the equation and parentheses. But if you take things step by step, you can find x with no problem.

First deal with those parentheses, using the distributive property. That means you need to *distribute* the 3 to each of the values inside the parentheses: $3 \times 2x + 3 \times 4 \rightarrow 6x + 12$. So now the equation looks like this: $6x + 12 = 4x + 10$. Now you need to get the x's on one side of the equation.

It makes more sense to subtract $4x$ from each side than to subtract $6x$ from each side. Why? Because that way you can avoid having a negative number multiplied by the variable: $2x + 12 = 10$.

Now get $2x$ by itself by subtracting 12 from each side of the equation: $2x = -2$. Divide each side by 2 to find out that x is -1.

What is the area of the semicircle found inside the unit square below? (The sides of a unit square measure 1 unit.

The area of a circle is $A = \pi r^2$.)

What is the area inside that square that is *not* also inside the semicircle?

The diameter of the semicircle is the same as the length of one side of the square. That means that the radius of the semicircle is ½ or 0.5. Now you can find the area of a circle with that radius: $A = \pi(0.5)^2 = 3.14 \times 0.25 = 0.785$ square units. But that's the area of a circle with a radius of 0.5. You need to take half to find the area of the semicircle: $0.785 \div 2 = 0.3925$ square units.

What about the space inside the square that doesn't include the semicircle? To find this area, you need both the area of the square and the area of the semicircle. Then you can subtract. The area of the square is 1^2 or 1. You found the area of the semicircle above, so the area of the rest of the space is 1 minus 0.3925 or 0.6075 square units.

Week Thirty-Eight

THURSDAY | APPLICATION

You're building a retaining wall that is 64 inches tall. You have two kinds of bricks. One is 8 inches thick and the other is 6 inches thick. How many rows of each kind of brick will you use? (Don't break the bricks into smaller ones.)

Remember that time when you thought math had only one right answer? Here's a counterexample, since there are several answers to this problem. Here you need to come up with the number of ways that 8s and 6s can be added together to give you 64.

The first answer might be pretty obvious. Sixty-four divided by 8 is 8, so you could just use 8 rows of 8-inch bricks and no 6-inch bricks. But what about combinations of 8-inch and 6-inch bricks? In this case, you're looking for multiples of 8 and 6 that add up to 64. (Since 64 is not evenly divisible by 6, you can't build the wall using only 6-inch bricks.)

Think about the multiples of 6: 6, 12, 18, 24, 30, 36, 42, 48, 54, and 60. And here are the multiples of 8: 8, 16, 24, 32, 40, 48, 56, and 64. You want one number from each list that when added together gives you 64. In fact, there are two pairs of these numbers: 24 and 40, and 48 and 16.

This doesn't mean that you want 24 rows of 6-inch bricks and 40 rows of 8-inch bricks. You have to figure out how many rows of 6-inch bricks fit in 24 inches, and how many rows of 8-inch bricks fit in 40 inches.

So, there are three possible answers to this problem: 8 rows of 8-inch bricks, 4 rows of 6-inch bricks plus 5 rows of 8-inch bricks, and 8 rows of 6-inch bricks plus 2 rows of 8-inch bricks. Which answer did you get?

Week Thirty-Eight

FRIDAY | PROBABILITY & STATISTICS

You roll a pair of six-sided dice. What is the probability of not rolling doubles?

What is the total probability of all of the possible outcomes? That would be 1. So the probability of *not* rolling doubles is 1 minus the probability of rolling doubles, right? This is called the *complement*.

So to find out the answer, you can find the probability of rolling doubles and then subtract from 1. List all of the possible doubles: (1, 1), (2, 2), (3, 3), (4, 4), (5, 5), and (6, 6). Now find the probability of rolling one double. To roll a (1, 1), you must roll a 1 and then roll another 1. The probability of rolling 1 is $1/6$. And since these are independent events, the probability of rolling another 1 is also $1/6$. So the probability of rolling (1, 1) is $1/6 \times 1/6 = 1/36$. There are six possible doubles, so the probability of rolling any of these doubles is $6/36$ or $1/6$.

But that wasn't the question, remember? You need to know the probability of not rolling doubles. Subtract the probability of rolling doubles from 1 to get $5/6$. This shows that there's a much greater chance of not rolling doubles than rolling doubles.

A teenager is caught sneaking out of the house. Her parents are angry and demand that she make a statement. They say: "If the statement is true, you will lose phone privileges for a week. If the statement is false, you must clean out the garage." Each of these punishments is equally horrible for this teen, so she gives her statement some thought. After hearing what their daughter has to say, the parents tell her that she has no punishment. What did the teen say?

The girl told her parents that she must clean out the garage. If that statement is true, then she loses her phone privileges for a week. But if she loses her phone privileges for a week, then she doesn't have to clean out the garage—and the statement she made is false.

Her parents cannot tell whether she is telling the truth or not, so they have no choice but to drop all threats of punishment.

Week Thirty-Eight

SUNDAY | GRAB BAG

There are 6 two-digit prime numbers. The sum of the digits of the individual numbers is divisible by 5. What are these numbers?

First off, you need to remember that a prime number is only divisible by 1 and the number itself. So 11 is a prime number, but 12 is not. Get it? Prime numbers are always odd, but not all odd numbers are prime. (For example, 27 is not prime, but 29 is.)

It's fairly easy to make a list of the two-digit prime numbers—or at least to start a list: 11, 13, 17, 19, 23, 29, 31, 37, . . . Take a look at this list. If you add the digits of any of these, is the result divisible by 5? Sure, there's 19: 1 + 9 = 10 and 10 ÷ 5 = 2. Then there's 23: 2 + 3 = 5 and 5 ÷ 5 = 1. And finally, there's 37: 3 + 7 = 10 ÷ 5 = 2.

So there are the first three of these numbers. But how can you find the remaining two? It might help to think of the two-digit numbers whose digits add up to 5, 10, or 15. (You can't get a higher multiple of 5 with two-digit numbers.) The numbers 41, 46, 50, 55, 64, 73, 82, and 91 fall into that category. Which ones are prime? You can eliminate all of the even numbers, which leaves only 41, 55, 73, and 91. But 55 is divisible by 11, and 91 is divisible by 7. The remaining numbers *are* prime: 41 and 73.

Two lists and the answer falls out.

Week Thirty-Nine

MONDAY │ NUMBER SENSE

How many 2s appear as digits in the numbers from 1 to 100?

This question is asking the number of times a 2 appears, when you list the numerals from 1 to 100. You can certainly list them and count. But there's a more elegant way of going about this. Think of the numbers in chunks, starting with 1 to 19. (The reason will become clear in a moment.) How many 2s are in that group of numbers? There's 2 and there's 12, so there are two 2s from 1 to 19. Now consider 20 to 29. There is one 2 in each tens place of these numbers, so there are ten 2s in the tens places. But there's also one 2 in the ones place (22), so the total number of 2s from 20 to 29 is 11. So now you're up to thirteen 2s.

Finally, look at the rest of the numbers: 30 to 100. In each grouping of 10, there is one 2. There are seven groupings of 10. So there are seven 2s from 30 to 100. Add that to 13, and you've got a total of twenty 2s in the numerals from 1 to 100.

TUESDAY | **ALGEBRA**

Solve for x.

$$4x - 40 = 7(-2x + 2)$$

You're probably getting used to the variables on both sides of the equation. But it's the parentheses that must be managed first. Distribute the 7 to each of the values inside the parentheses: $4x - 40 = -14x + 14$. Now you can start solving for x. Add $14x$ to both sides of the equation. This leaves 14 by itself on the right side of the equation: $18x - 40 = 14$. Now add 40 to both sides, which undoes the subtraction on the left side: $18x = 54$. You'll find out that $x = 3$, when you divide each side by 18.

Week Thirty-Nine

WEDNESDAY | **GEOMETRY**

What is the length of the missing side?

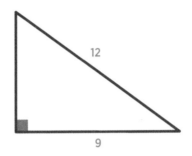

First identify the hypotenuse, or *c*. It's always the longest side and always opposite the right angle. So *c* = 12. Now you can substitute and solve for the variable: $(9)^2 + b^2 = (12)^2$ ➔ $81 + b^2 = 144$ ➔ $b^2 = 63$ ➔ $b = \sqrt{63}$. You can leave your answer like this or use a calculator to find the decimal equivalent (7.94).

Notice that the lengths of the sides are not all whole numbers. So this triangle is not an example of a Pythagorean triple.

THURSDAY | APPLICATION

Barbara knows she should be putting 10% to 20% of her monthly income into her savings account each month. She takes home $4,325 per month and puts away $750. Has she met her goal? What percentage is she putting away?

You can find 10% and 20%, and see if the amount is between the two. And this is a really simple process. To find 10% of $4,325, move the decimal point one place to the left: $432.50. Double this to find 20%: $432.50 × 2 = $865. Looks like Barbara is good to go.

But what percentage is she putting away? This is the second way to find out if Barbara is in the zone. Divide the amount she is putting away by her monthly income: $750 ÷ $4,325 = 0.17 or 17%, which is between 10% and 20%.

Week Thirty-Nine

FRIDAY | PROBABILITY & STATISTICS

Odds are the ratio of the probability of an event happening to the probability of that event not happening. The probability of your football team winning next week is $\frac{5}{8}$. What are the odds? Also, if the odds of your team winning are 5:4, what is the probability?

If the probability of your team winning is $\frac{5}{8}$, the probability of your team not winning is $1 - \frac{5}{8}$ or $\frac{3}{8}$. So the odds of your team winning are $\frac{5}{8} : \frac{3}{8}$ or 5:3.

If the odds of your team winning are 5:4, then the probability of your team winning is 5 over 5 + 4, or $\frac{5}{9}$.

Week Thirty-Nine

SATURDAY | LOGIC

There are four cards lying on a table. The first card has a picture of a fish. The second card has a 4 on it. The third card has a 13 on it, and the fourth card has a picture of a cat. You are told that when there is a picture of a mammal on the card, the opposite side has an odd number. Which cards do you flip to test this rule? (See if you can do this in the least number of card flips.)

Of course you can flip all of the cards over to see what happens. But there is a way to do this in only two flips. If you flip the cards with the animals, you won't know if the rule holds for the cards with the numbers. Likewise, if you flip the cards with the numbers, you won't know if the rule holds for the cards with the animals. So it's best to flip one card with an animal and one with a number. This leaves two logical choices. Here's the first: the first card (with the fish) and the second card (with the 4). If the rule holds, the fish card will have an even number, and the 4 card will have a picture of a non-mammal. The second option is to flip the 13 card and the cat card. If the rule holds, the 13 card will have a picture of a mammal, and the cat card will have an odd number.

Week Thirty-Nine

SUNDAY | GRAB BAG

A hotel has 1,000 rooms. Each room is numbered from 1 to 1,000. How many 9s are there in the room numbers?

This is a counting problem, and it helps to look at the single-digit numbers first. In the digits 1 through 9, there is one 9. The same is true for all of the sets of tens until 90. That makes nine 9s in the numbers from 1 to 89. There are eleven 9s in the numbers from 90 to 99. So from 1 to 99, there are nineteen 9s.

The same is true for 100 to 199, 200 to 299, . . . 800 to 899. That brings the total to 171 (9 × 19 = 171). But 900 to 999 is a little tougher. Each of these numbers has at least 1 nine in the hundreds place. That's an extra 100 nines, for a total of 271 nines. Then there are the 10 nines in the tens place, and we're up to 281 nines. Finally, count the nines in the ones place—there are 19. And that means there are three hundred 9s in the numbers from 1 through 1,000.

53 is 27% of what number?

You're being asked to find the whole amount, given the part. Like before, there are two ways to solve this problem. First, you can translate the words in the question. "Is" means *equal*, "of" means *multiplication*, and "what number" means x. So the question can be written as an algebraic equation this way: $53 = 27\% \times x$. First change the percentage to a decimal: $53 = 0.27x$. Now just divide each side of the equation by 0.27, and you'll have $x = 196.3$ (rounded to the nearest tenth).

But what about the proportions approach? It works here, too. In this case, 53 is the part, not the whole, so the proportion will look like this: $\frac{53}{x} = \frac{27}{100}$. Cross multiply to get $5,300 = 27x$. And divide each side of the equation by 27 to get the x by itself. You'll get the same answer as above. How about that?

TUESDAY | ALGEBRA

Factor the following expression:

$$8x + 12$$

This problem is simple, as long as you remember what a factor is. A *factor* is a number that divides evenly into another number. You can factor out a 4 from both $8x$ and 12, which gives you $4(2x + 3)$.

Notice that if you use the distributive property, you'll get the original expression. That's how you know you have two factors.

WEDNESDAY | GEOMETRY

△ABC ≅ △XYZ. (Remember, that's a congruence symbol between the two triangles.) Write six congruence statements for each of the corresponding pairs of sides and angles. (Three of these statements will have sides, and three will have angles.)

The trick to writing congruence statements for congruent triangles is to watch the order of the letters.

In the triangle congruence statement above—△ABC ≅ △XYZ—the *A* corresponds to the *X*, the *B* corresponds to the *Y*, and the *C* corresponds to the *Z*. This makes it very easy to write congruence statements for the angles:

∠A ≅ ∠X, ∠B ≅ ∠Y, and ∠C ≅ ∠Z.

Do the same thing for the sides, naming each by their endpoints: $\overline{AB}$ ≅ $\overline{XY}$, $\overline{BC}$ ≅ $\overline{YZ}$, and $\overline{AC}$ ≅ $\overline{XZ}$.

Luckily, math folks like to use alphabetical letters in naming figures, which makes the process of writing corresponding congruent parts much easier.

Week Forty

You have had your eye on a new set of cookware at the fancy-schmantsy home-goods store. Your budget is no more than $125, but these babies are pricey—selling for $195. You just saw that the store is having a 30%-off sale. Can you finally afford the cookware set?

In this situation, you don't need an exact value for the cookware. You just need to have an idea of whether or not the sale price will be in your budget. So you can round, making the math easier.

Round the original price up to $200, and then find 30% of this. You can use proportions or just multiply 200 by 0.3. Or you can notice something interesting here: $200 is twice as much as $100. Thirty percent of $100 is $30, so what is 30% of $200? (That would be $60, chef.)

No matter how you do the math, you find out that the sale price is $60 off of the regular price. When you subtract $60 from $200, you get $140, which is still not quite at your budget. You'll need to wait a little longer.

Want to do the problem in one step? Just multiply $200 by 0.7 to get right to the answer. It's not really a one-step problem, since you're subtracting 0.3 from 1 to get 0.7.

(P.S. Would it make a difference if you found 30% of the exact price, $195? In this case, nope. Thirty percent of $195 is $58.50. So the exact sale price is $136.50, still over budget.)

FRIDAY | **PROBABILITY & STATISTICS**

The bar graph below shows the sales at a bakery by item. If the bar graph were changed to a circle graph, how many degrees would the chocolate chip muffin wedge be?

Bakery Sales

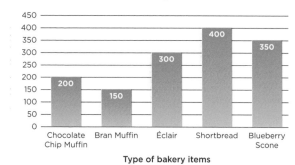

Type of bakery items

So here's the thing about circle graphs: each wedge must be correctly proportional to the whole. Find the percentage of bakery items that are chocolate chip muffins, and you'll then be able to find the size of that wedge in a circle graph.

To find the percentage of sales that were chocolate chip muffins, divide the number of chocolate chip muffins by the total items sold. First add the total of each of the pastries: 200 + 150 + 300 + 400 + 350 = 1,400. Now divide: 200 ÷ 1400 = 0.14 or 14%.

Next apply this percentage to the circle graph. Each circle has a total of 360°. What is 14% of 360? Multiply to find out: 0.14 × 360 = 50.4. So the wedge must be 50.4°.

Notice that this wedge is not 50% of the circle. It's only 14%, which is 50.4°.

SATURDAY | LOGIC

Julianna is a manager at a widget and thingamabob factory. In front of her, she has three boxes, but *all of them* are mislabeled. One box contains only widgets, another contains only thingamabobs, and the third crate contains both widgets and thingamabobs. Each crate has one of the following labels: "widgets," "thingamabobs," and "widgets & thingamabobs." Julianna opens one box and takes out one item. She then then declares that she knows exactly what is in each box. How did she do that?

Julianna might *seem* clairvoyant, but she's not. She just used her noggin to help solve this problem—which is why she's paid the big bucks, of course. The biggest clue? *All* of the boxes are mislabeled.

Julianna chose to take an item from the box marked "widgets & thingamabobs." If it is a widget, she knows the box should be labeled "widgets." This means the box marked "widgets" should be labeled "thingamabobs," and the box marked "thingamabobs" should be labeled "widgets & thingamabobs."

Week Forty

SUNDAY | GRAB BAG

What is the square of 111,111,111?

Sure, you can plug this number into a giant scientific calculator, but what's the fun in that? Turns out there's an interesting pattern to be discovered. Start by squaring 11: $11^2 = 121$, which is like a numerical palindrome. Does that continue for 111^2? Yep. $111^2 = 12,321$. Another palindrome.

Keep going to see if there's a pattern: $1111^2 = 1,234,321$ and $11111^2 = 123,454,321$. These are all palindromes beginning with 1. The largest digit is the number of digits in the number being squared. There are four digits in 1111, so $1111^2 = 1,234,321$.

How many digits are there in 111,111,111? There are 9, so $111,111,111^2 = 12,345,678,987,654,321$. Pretty cool, eh?

7 is what percentage of 123?

For this problem, you want the percentage. You can definitely translate the question into an algebraic equation: $7 = x \times 123$ or $7 = 123x$. Divide each side of the equation by 123 to isolate the variable: 0.06 (rounded to the nearest hundredth). Next change the decimal to a percentage, by moving the decimal point two places to the left: 6%.

But if you can't remember how to set up the algebraic equation (or which number to divide by), you are more than welcome to use proportions to find the answer. Remember to keep the parts together and the wholes together. The 7 is part of the 123, so your first fraction is $^7/_{123}$. You don't know the percentage, but you do know that it's part of 100. This means your second fraction is $^x/_{100}$. Write the proportion, cross multiply, and then solve for x: $^7/_{123} = {}^x/_{100}$ ➜ $700 = 123x$ ➜ $x = 6$ or— ready for it?—6%.

Notice that you don't need to move the decimal point when you use proportions. That's because the percentage is already in the proportion, as the denominator of one of the ratios.

Week Forty-One

TUESDAY | **ALGEBRA**

The FOIL method is used to multiply two binomials, like $(x + 8)(x + 1)$.

In this process, you multiply the First terms (x and x), the Outer terms (x and 1), the Inner terms (8 and x), and the Last terms (8 and 1).

Use FOIL to multiply $(x + 8)(x + 1)$. Then combine like terms to simplify the expression.

Start by multiplying the first two terms: $x \times x = x^2$. Then multiply the outer terms: $x \times 1 = x$. Next up are the inner terms: $8 \times x = 8x$. And finally the last terms: $8 \times 1 = 8$. Put it all together to get $x^2 + x + 8x + 8$.

But this expression is not in its simplest form. There are two like terms: x and $8x$. Because each of these has an x, you can add them. And $x^2 + 9x + 8$ is the final answer.

Week Forty-One

WEDNESDAY | GEOMETRY

How many squares are in this drawing?

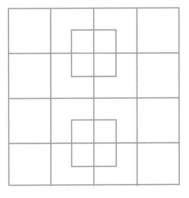

The big deal here are the overlapping squares, but start with the easy ones. Looking at the largest square, there are 4 squares in each row and 4 squares in each column, which is 16 squares total. Then there are the 2 squares that overlap the middle two columns. Those 2 squares are divided into 4 squares, so we're looking at 10 squares in all. The total is now 26. Now for the overlapping squares. Looking at the squares row by row, there are 3 overlapping 2 × 2 squares in each of rows one and two, rows two and three, and rows three and four.

That makes 9 more squares, which gives 26 + 9 or 35. What about the 3 × 3 squares? There are 2 in rows one, two, and three and 2 in rows two, three, and four, or 4 in all. The total is now 39, and all you have left to count is the largest square of all, the 4 × 4. Therefore, there are 40 squares in this picture. Is that more—or less—than what you thought?

Twenty percent of a city's bus fleet runs on natural gas. The city wants to replace buses until 50% run on natural gas. If there are 750 buses in the fleet, how many more regular gas buses must be replaced to reach 50%?

Start by answering this question: How many buses currently run on natural gas? This is simple to discover. Just find 20% of 750: 0.20 × 750 = 150. So there are 150 buses that currently run on natural gas. Next, find out 50% of 750: 0.50 × 750 = 375. That's the number of natural gas buses the fleet will have when it reaches its goal of 50% natural gas buses.

Now, subtract: 375 – 150 = 225. So the fleet needs to replace 225 of regular gas buses with natural gas buses to reach its goal of 50%.

Week Forty-One

FRIDAY | PROBABILITY & STATISTICS

You roll a pair of six-sided dice. What is the probability that the sum of the outcomes is either 8 or 10?

Start by listing the possible outcomes that add up to 8 and 10:

Sum of 8 ➜ (2, 6), (3, 5), (4, 4), (5, 2), (6, 2)

Sum of 10 ➜ (4, 6), (5, 5), (6, 4)

So there are 5 possible outcomes with a sum of 8 and 3 possible outcomes with a sum of 10.

How many total outcomes are there? Since there are 6 possible outcomes for each die, there are 36 total outcomes. That means there's a $5/36$ probability of rolling a sum of 8 and a $3/36$ probability of rolling a sum of 10.

But you want a sum of 8 or a sum of 10. To find the answer, add the two probabilities: $5/36 + 3/36$. This means there is an $8/36$ or $2/9$ probability of rolling an 8 or a 10.

Week Forty-One

SATURDAY | LOGIC

A cube with 5-inch sides is painted blue. Then the cube is cut into 1-inch cubes. Each of these smaller cubes will have 3 painted sides, 2 painted sides, 1 painted side, or no painted sides. How many of each will there be?

It might help to draw a picture for this one. First, find out how many 1-inch cubes there are in all. Multiply 5 × 5 × 5 to find out that there are 125 smaller cubes.

Think about the first option: 3 painted sides. Where will these cubes have to be in the larger cube? They must be on the corners of the cube, right? There are 8 corners, so there are 8 cubes with 3 painted sides.

Next up are the cubes with 2 painted sides. These must be along the edges of the cube—but not the corners. Since there are 5 cubes along each edge, and 2 of them are corners, there must be 3 cubes along each edge that are not corners. And since there are 12 edges, there are 12 × 3 or 36 cubes with 2 painted sides.

Now for the cubes with 1 painted side. These are cubes that are on the outside of the larger cube (the faces), but not on an edge or corner. On each face, there are 3 × 3 or 9 of these cubes. There are 6 faces, so there are 9 × 6 or 54 cubes with 1 painted side.

To find the cubes with no painted sides, subtract the total cubes with at least 1 painted side from the total number of cubes: 125 – (8 + 36 + 54) = 125 – 98 = 27. This makes sense. If you remove all of the outside cubes, you get a smaller, 3 × 3 × 3 cube, which is made up of 27 one-inch cubes.

SUNDAY | **GRAB BAG**

A mother shares a box of candy with her children. Her favorite child gets $\frac{1}{2}$ of the candies. Her second favorite child gets $\frac{1}{4}$ of the candies. Her next favorite gets $\frac{1}{5}$ of the candies, and her least favorite gets the rest, which is 3 pieces. How many pieces of candy are in the box?

O ne way to solve this problem is by creating an algebraic equation. But first, since the fractions are simple, change them to decimals:
$\frac{1}{2}$ = 0.5, $\frac{1}{4}$ = 0.25, and $\frac{1}{5}$ = 0.2.

If you let x be the number of candies, you can set up an expression: $0.5x + 0.25x + 0.2x$. Then there's the poor unfavored child: $0.5x + 0.25x + 0.2x + 3$. Now you can make an equation, since all of these add up to the total number of candies (which is x): $0.5x + 0.25x + 0.2x + 3 = x$.

Combine all of the x's to get $0.95x$: $0.95x + 3 = x$. Now solve for x by subtracting $0.95x$ from both sides and then dividing: $3 = 0.05x \rightarrow 60 = x$.

So there are 60 pieces of candy in the box. The favorite child gets 30 pieces, the second favorite gets 15 candies, the third favorite child gets 12, and that last poor child only gets 3 pieces.

Week Forty-Two

MONDAY | NUMBER SENSE

$$-81 \div 9 = ?$$

Okay, so you already know what $81 \div 9$ is, right? But what happens when one of these numbers is negative? Just like with multiplication, when you divide numbers with opposite signs, the answer is always negative.

What if you can't remember that rule? Using a number line is a little trickier, but it does work. In this case, you want a number line that goes from -81 to 81. Just like with multiplication, start at 0. Count by 9s until you get to -81. You took 9 jumps, and because you're going to the left, your answer is negative. Therefore, the correct answer is -9.

TUESDAY | ALGEBRA

Use FOIL (first, outer, inner, last) to multiply $(x - 3)(x + 1)$. Then combine like terms to simplify the expression.

Remember that FOIL refers to the order in which you multiply each of the terms of the binomials. (By the way, do you know why these are called *binomials*? There are two (*bi*) terms in each expression.)

F is x^2; O is x; I is $-3x$ and L is -3. Notice that there are some negative numbers in this problem. Adding a negative is the same as subtracting a positive, so turn those bad boys into subtraction when you write the expression: $x^2 + x - 3x - 3$. Now simplify, paying close attention to the signs when you combine like terms: $x^2 - 2x - 3$. The binomials are multiplied to form a polynomial. (*Poly* means "many," of course.)

Week Forty-Two

WEDNESDAY | **GEOMETRY**

Find the measures of the missing angles:

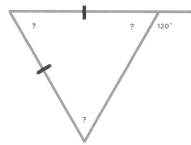

The first thing you might notice is that the 120° angle is supplementary to the top right angle of the triangle. That is, the sum of these two angles is 180°. So the measure of the top right angle is 180° – 120° or 60°. But how can you find the measures of the other two angles? Those little lines on the top and left sides indicate that the sides have the same measure. That means the triangle is an isosceles triangle. Base angles of an isosceles triangle have the same measure. So the bottom angle is also 60°. And that leaves only one option for the third angle: 60°. The sum of the measures of the angles of a triangle is 180°.

In fact, you've uncovered something interesting about this triangle—not only is it isosceles, but it's equilateral.

Week Forty-Two

THURSDAY | APPLICATION

Gertrude is twice as old as Alice, who is $^2/_3$ of Virginia's age. If Alice is 36 years old, how old are Gertrude and Virginia?

Since Gertrude is twice as old as Alice, she's 2 × 36 or 72 years old. Alice is $^2/_3$ as old as Virginia, which means Virginia is $^3/_2$ as old as Alice. (Here's one way to think of this: Alice = $^2/_3$ × Virginia → Alice ÷ $^2/_3$ = Virginia → Alice × $^3/_2$ = Virginia.) So to find Virginia's age, multiply 36 by $^3/_2$: 36 × $^3/_2$ = $^{108}/_2$ = 54 years old.

Week Forty-Two

FRIDAY | PROBABILITY & STATISTICS

You roll two six-sided dice. What is the probability that the sum of the dots will be either an even number or less than 6?

There are 6 sides on each die, so the total number of outcomes is 6 x 6 or 36. The probability will be the outcomes that meet the criteria over the total possible outcomes.

But which outcomes meet the criteria? This is where a list can be helpful. To roll an even number, you'll have one of the following outcomes: (1, 1), (1, 3), (1, 5), (2, 2), (2, 4), (2, 6), (3, 1), (3, 3), (3, 5), (4, 2), (4, 4), (4, 6), (5, 1), (5, 3), (5, 5), (6, 2), (6, 4), or (6, 6).

To roll a value that's less than 6, you'll have one of the following outcomes: (1, 1), (1, 2), (1, 3), (1, 4), (2, 1), (2, 3), (3, 1), (3, 2), or (4, 1).

You might have noticed that there are some values listed in each list. These would work in an *and* situation, but for an *or* situation, only one of them can stay in the lists. So in the even list, cross out any values that also occur in the less-than-6 list. Then combine the lists, leaving the following: (1, 1), (1, 2), (1, 3), (1, 4), (1, 5), (2, 1), (2, 2), (2, 3), (2, 4), (2, 6), (3, 1), (3, 3), (3, 5), (4, 1), (4, 2), (4, 4), (4, 6), (5, 1), (5, 3), (5, 5), (6, 2), (6, 4), (6, 6). Everything in this list fits one of the two criteria: the sum is even or the sum is less than 6. There are 24 of these options, so the probability is $^{24}/_{36}$, which can be reduced to $^2/_3$.

Week Forty-Two

SATURDAY | LOGIC

You have a bunch of flowers and a set of vases. If you put 4 flowers into each vase, 1 vase will still be empty. If you put 3 flowers into each vase, 1 flower will be left over. How many flowers and vases do you have?

In short, you're looking for a number that is 3 times plus 1 of another number. If there are 16 flowers, you can divide the flowers into 4 vases. That means there are 5 vases. But does this work for the second option? Putting 3 flowers into 5 vases gives you 15 flowers. That leaves 1 flower left over.

Jack's mother was 36 when he was born. How old is Jack when his age is the reverse of his mother's?

There are actually several answers to this problem, depending on how you interpret the question and the numbers involved. Start by thinking about the place values of the numbers you're trying to find.

The ones place of the mother's age will need to become the tens place of Jack's age. You could interpret the question this way: When is the *first* time that Jack's age is the reverse of his mothers? And then again, you have a decision to make. Will you count ages that have a 0 in the tens place? If so, the first time this happens is when Jake is 4 and his mother is 40: 40 – 36 = 04.

But if you don't count that solution, the first time is when his mother is 51 and Jake is 15. Now, you might be noticing a possible pattern. The next time this happens is when the tens place is one greater and the ones place is one greater—that's 62 and 26. This pattern continues on and on.

Week Forty-Three

MONDAY | NUMBER SENSE

$$4 + 6^2 \div 9 = ?$$

This time, you've got to add, divide, and square a number. But in what order? Going from left to right isn't going to get you to the correct answer. Instead, you need to follow the order of operations: Please Excuse My Dear Aunt Sally. The E for "excuse" stands for *exponent*, so start there: $4 + 36 \div 9$. Now divide to get: $4 + 4$. Finally, just add to get 8.

What would have happened if you had ignored the order of operations? $4 + 6 = 10$. Raising 10 to the second power gives 100, and then dividing by 9 is a little more than 11. So nope. Not the same answer. Follow the order of operations, and you'll be in good shape.

Use FOIL to multiply $(x + 1)(x - 1)$.
Then simplify by combining like terms.

Once you know that FOIL stands for first, outer, inner, last, this process becomes pretty simple. Of course, you always need to watch out for the signs (or subtraction). In this case, F is x^2; O is $-x$; I is x; and L is -1.

Now to build and simplify the expression: $x^2 - x + x - 1$. But how do you simplify $-x + x$? When you add $-x$ to x, you get 0, so $x^2 - 1$ is the simplified expression.

In this case, multiplying two binomials results in a binomial. In fact, it's a special kind of binomial, called a difference of squares. (Both x^2 and 1 are perfect squares.)

Week Forty-Three

WEDNESDAY | GEOMETRY

In the drawing below, each small box measures 1.5 units by 1.5 units by 1.5 units. What is the volume of the figure?

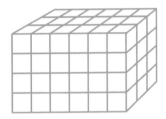

The dimensions of the figure are 6 boxes by 4 boxes by 3 boxes. Therefore, there are 6 × 4 × 3 or 72 boxes in all. The boxes measure 1.5 × 1.5 × 1.5, so the volume of each box is 3.375. The volume of the figure is 243 cubic units. You can find this by multiplying the number of boxes by the volume of each box.

Week Forty-Three

THURSDAY | APPLICATION

Jamal has a green thumb but only a very tiny yard to plant his vegetable garden. Given the varieties of tomatoes, cucumbers, peppers, and herbs that he wants to plant, he needs a rectangular space with a perimeter of 16 feet. The longest side he can manage is 5 feet. How long must each of the other sides be? (Sketch a picture, if you need to.)

Since Jamal's garden is a rectangle, you know that the opposite sides must have the same length. (Why? Well, that's in the definition of a rectangle.) So if one side measures 5 feet, the side opposite it must also measure 5 feet. But what about the remaining sides?

The perimeter—or the distance around the whole garden—is 16 feet. Add the two sides that you know: 5 + 5 = 10.

Then subtract from 16: 16 – 10 = 6. This gives the total of the two missing sides. Since these sides must be equal in length, you can simply divide 6 in half to find the length of each side: 6 ÷ 2 = 3. So the garden must be 5 feet × 3 feet.

Week Forty-Three

FRIDAY | PROBABILITY & STATISTICS

A couple has 2 children, and 1 is a girl. Assume that the probability of each gender is ½. What is the probability that the couple's other child is a girl?

Did you first guess ½? That's a common—but incorrect—answer. Instead, you need to consider the possible outcomes. The couple can have 2 girls (GG), a girl and a boy (GB), a boy and a girl (BG), or 2 boys (BB). Since the couple already has 1 girl, the probability of having another—or 2 girls—is 1 out of 4.

$$\frac{A}{C} = \frac{B}{F} = \frac{E}{O} = \frac{?}{X}$$

What is the missing letter?

The first thing you should know about this problem is that it's not really about letters. The equal signs are a big clue of this, since only numbers—not letters—can truly be equal. So you're actually looking for equivalent fractions that can be represented by the letters.

Notice that the letters in the numerators are in alphabetical order. *A* comes before *B*, which comes before *E*. The same is true for the denominators. If you replace the letters with their numerical place in the alphabet, you will get really close to the answer. Since *A* is the first letter of the alphabet, it becomes 1. *B* is 2, and *E* is 5. In the denominators, you have 3, 6, and 15. Noticing a pattern yet?

With this rule, $^A/_C$ becomes $1/3$, $^B/_F$ becomes $2/6$, and $^E/_O$ becomes $5/15$. What would that make the last fraction in numerical terms? To make each of the fractions, you multiply by a number over itself: $1/3 \times 2/2 = 2/6$ and so on. Therefore, the last fraction is $6/24$. In fact, *X* is the 24th letter of the alphabet. What is the 6th? *F,* which is the missing numerator.

Week Forty-Three

SUNDAY | GRAB BAG

The sum of two consecutive *even* integers is 26.
What are the two numbers?

onsecutive even integers are the pairs of numbers when you count by 2: -6, -4, -2, 0, 2, 4, 6, 8, 10, . . . In this list, 2 and 4 are consecutive even numbers.

Can one or both of these numbers be negative? Both cannot be negative. When you add two negative numbers, you get a negative number. But when you add a positive and a negative number, you can get a positive number.

So for sure, you're going to have two positive numbers. You can certainly approach this problem by guess and check. If you extend the list of even numbers, you can pull out the two consecutive even numbers that add up to 26: 2, 4, 6, 8, 10, 12, 14, 16, 18, 20, 22, 24, 26. Why stop at 26? You could have stopped earlier, actually, because 26 + 0 = 26, and 26 and 0 are not consecutive even numbers. Looking at this list, which of the consecutive even numbers add up to 26? That would be 12 and 14.

You can also approach this problem algebraically. Let the first integer be $2x$ and the second integer be $(2x + 2)$. Then you can write and solve the equation $2x + (2x + 2) = 26$ → $4x + 2 = 26$ → $4x = 24$ → $x = 6$. If $x = 6$, then the first number is 2(6) or 12. The second number is 2(6) + 2 or 14. See? Algebra can come in handy.

Week Forty-Four

MONDAY | **NUMBER SENSE**

The Greatest Common Factor (GCF) is the largest number that will divide evenly into a set of numbers. What is the GCF of 76 and 24?

You can certainly list all of the factors of 76 and 24 (or the numbers that divide evenly into each number). But 76 is a pretty large number, so being smart with trial and error might be more efficient.

With that in mind, the first number you should check is 24. Why? Well, 24 is a factor of itself—in other words, 24 does divide evenly into 24. And since 76 is an even number, there's a slight chance that 24 will divide evenly into 76. Unfortunately, that's a no-go: $76 \div 24 \approx 3.167$.

You'll need to try another route. How about 12? $24 \div 12$ is 2, which makes 12 a pretty big factor of 24. This choice doesn't work either: $76 \div 12 \approx 6.333$.

Now you're down to 8, the next largest factor of 24. It's not a factor of 76, and neither is 6, the next largest factor of 24. That leaves 4, which is a factor of 76. So 4 is the GCF of 24 and 76.

Week Forty-Four

TUESDAY | ALGEBRA

Use FOIL to multiply $(x - 2)(x - 3)$. Then simplify by combining like terms.

FOIL is a pretty simple process, but in this problem there are two negative numbers. It's a good idea to pay *really* close attention to the signs.

F is x^2; O is $-3x$; I is $-2x$; and L is 6. (Remember that when you multiply two negatives, like -2 and -3, you get a positive, 6.)

Put it all together: $x^2 - 3x - 2x + 6$. And simplify to get $x^2 - 5x + 6$. Notice that $-3x - 2x$ is equal to $-5x$. In other words, you add and make the answer negative.

Week Forty-Four

WEDNESDAY | **GEOMETRY**

A *transformation* moves a point or a set of points from one place on the Cartesian plane to another. One kind of transformation is a *translation* or *slide*. The graph below shows that point *A* (2, 2) was translated or slid to point *B* (4, -3).

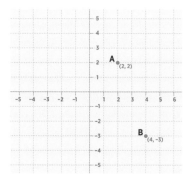

How many units was the point moved in any direction?
(Choose the least number of movements.)

Given the location of point *B*, it looks like point *A* was moved to the right and down. If you count the horizontal units between *A* and *B*, you get 2 units. And if you count the vertical units between these points, you get 5 units. So point *A* was slid 2 units to the right and 5 units down.

Another way to look at this is by considering the coordinate pairs for each of the points. If you subtract the *x* value of *A* from the *x* value of *B*, you get 2. Since this is the difference between *x*-values and is positive, the direction is to the right. Subtract the *y*-value of *A* from the *y*-value of *B* to get -5. Since this is the difference between *y*-values and is negative, the direction is down.

Week Forty-Four

THURSDAY | APPLICATION

A Little League team won 40% of its games, or 8 games. How many games did the team play in all?

Hold up! Before you find 40% of 8, take another look. You know that 8 is 40% of the number of games played, so this is a different question. Here's what's being asked: 8 is 40% of what number?

If you can remember how to set up an equation to solve, go for it. But most of us have other things jamming our already-crowded brains, so a proportion comes in handy. In case you've forgotten, here's a definition: a proportion is two equal fractions (or ratios). One of these fractions will represent the percentage: $^{40}/_{100}$. The other will represent the part over the whole. Since you don't know the whole, that's where your variable will go: $^{8}/_{x}$. This gives you the following proportion:

$$\frac{40}{100} = \frac{8}{x}$$

Now you can cross multiply and solve for x: $40x = 800 \rightarrow x = 20$. So there were 20 games in all.

Notice that if you simply divide 8 by 0.4, you also get the correct answer. Choose the method that makes most sense to you.

FRIDAY | PROBABILITY & STATISTICS

A television executive is scheduling six 30-minute programs for primetime Monday night. In how many different ways can these programs be scheduled?

If you don't know the rule for this, you can simply identify each of the programs—use letters to keep everything neat and tidy—and then create a tree diagram that will connect each show with the time slots that are available. But there is a pretty slick way to find the answer without a tree diagram.

When you place one of the shows in the first slot, there are only 5 options for the remaining 5 shows. And then there are only 4 options for the 4 remaining shows, and so on, until you get to 1. This is a *factorial*, written with an exclamation point: 6!

This translates to $6 \times 5 \times 4 \times 3 \times 2 \times 1$. So there are 720 different ways that these shows can be scheduled. That's a lot more options than anyone needs.

I was 27 the day before yesterday, and next year I'll be 30.
When is my birthday?

Whew! That's a lot of time passing in a year.
The only way to make this problem work is if my
birthday is on December 31. On December 30, I was 27.
On January 1, the day we're talking about, I am 28.
On December 31 of this year, I'll be 29. So next year,
I'll be 30.

Week Forty-Four

SUNDAY | GRAB BAG

A *Pythagorean triple* is a set of three positive integers that fits the Pythagorean theorem: $a^2 + b^2 = c^2$. The variables represent the lengths of sides of a right triangle, where a and b are the shortest sides (the legs) and c is the longest side (the hypotenuse). Can you list three sets of Pythagorean triples?

You are looking for two squares that, when added together, give another square. One obvious answer is 1, 0, and 1. But the side of a triangle cannot measure 0. So, you're going to have to go back to the drawing board.

To start, how about making a list of perfect squares? 1, 4, 9, 16, 25, 36, 49, 64, 81, 100, 121, 144, 169, . . . Do any of these fit the Pythagorean theorem? Yep. Take a look at 9, 16, and 25: 9 + 16 = 25. This means that 3, 4, and 5 are a Pythagorean triple. And there's one more: 25 + 144 = 169. So, 5, 12, and 13 are another Pythagorean triple.

But that's where this list runs dry. Look at the triples that you've already found. Notice the pattern? The first number, a, is always odd. The second number, b, is the square of a minus 1 and then divided by 2: $(3^2 - 1) \div 2 = (9 - 1) \div 2 = 8 \div 2 = 4$ and $(5^2 - 1) \div 2 = (25 - 1) \div 2 = 24 \div 2 = 12$. (Cool, huh?) Finally, the last number, c, is b plus 1: 4 + 1 = 5 and 12 + 1 = 13.

With that in mind, try the next odd number: 7. Let $a = 7$, which means that $b = (7^2 - 1) \div 2$, and $b = 24$. So $c = 24 + 1 = 25$. Do the numbers 7, 24, and 25 fit the Pythagorean theorem? $7^2 + 24^2 = 25^2 \rightarrow 49 + 576 = 625 \rightarrow 625 = 625$. So that's the next Pythagorean triple.

Week Forty-Five

MONDAY | NUMBER SENSE

Which of the following gives the largest answer?

$$3 + 4 + 5 = ?$$

$$1 \times 2 \times 6 \times 1 = ?$$

$$1 \times 3 + 2 \times 2 + 5 \times 1 = ?$$

The straightforward way to come to this answer is to work out each of the problems, one by one. In the first, you only need to add: $3 + 4 + 5 = 7 + 5 = 12$. In the second, there is only multiplication: $1 \times 2 \times 6 \times 1 = 2 \times 6 \times 1 = 12 \times 1 = 12$.

The third problem contains both addition and multiplication, so you need to think about the order of operations. Start with multiplication: $(1 \times 3) + (2 \times 2) + (5 \times 1) = 3 + 4 + 5$. That's exactly the first problem. And so all three have the same answer, and none is greater than the others.

TUESDAY | ALGEBRA

$(x + 2)^2$?

When you square a number, you multiply it by itself. Same with a binomial—or any algebraic expression. So $(x + 2)^2 = (x + 2)(x + 2)$. Now you can multiply, using FOIL. F is x^2, O is $2x$, I is $2x$, and L is 4. Put all of these together using addition to get $x^2 + 2x + 2x + 4$. Simplify by combining like terms: $x^2 + 4x + 4$.

Week Forty-Five

WEDNESDAY | GEOMETRY

A point was translated (slid) from (1, 0) to (5, 0). Without graphing these points, what can you say about how this point was translated?

Since you know the point was translated, you know that the difference between the x-values indicates how the point was slid horizontally. And the difference between the y-values indicates how the point was slid vertically. So the point was slid to the right 4 units. But it was not slid vertically at all.

Valerie is painting her bedroom. The ceilings are 8 feet tall, and the room itself is 15 feet by 12 feet. Each gallon of paint will cover 350 square feet. She wants to use 2 coats of paint for each wall, and she's not painting the floor or ceiling. She's also decided not to take into consideration the windows or doors. (She's going for a close approximation.) How much paint should she buy?

This is actually a surface area problem. Once Valerie finds the area of the space being painted, she can divide by 350 to find the number of gallons of paint she'll need.

She's painting 4 walls. Two of the walls are 8 feet by 15 feet, and the other 2 walls are 8 feet by 12 feet. Find the area of each of the walls and then add them together to get the surface area: 8 × 15 = 120 and 8 × 12 = 96. Double each of these, since there are 2 walls of each size: 120 × 2 = 240 and 96 × 2 = 192.

Now add: 240 + 192 = 432. Valerie needs to cover 432 square feet. But remember, she wants to use 2 coats of paint: 432 × 2 = 864 square feet. Divide by 350 to get the number of gallons needed: 864 ÷ 350 ≈ 2.5 gallons of paint. (It's actually 2.46857. . . but it makes sense to round to 2.5.)

Of course you can also switch up the steps in this problem. For example, find find the number of gallons of paint needed to paint 1 coat, and then double that amount. But whatever you do, make sure to follow the order of operations.

FRIDAY | **PROBABILITY & STATISTICS**

Cousin Bubba loves to play the craps tables. His plan is simple: he avoids the hot tables—tables where folks are winning—and looks for losing streaks. At a losing table, he places his bet on 7 or 11 and waits to win big time. What do you think of Bubba's plan?

Poor Cousin Bubba is falling victim to the gambler's fallacy. But he's not alone in this misconception. Casinos count on its patrons to make the same assumption.

It's human nature to assume that the probability of a win is higher when there's been a losing streak. It's time for luck to turn around, right?

The problem with this theory is that rolling a pair of dice is always an independent event. Think of it as starting over. No matter how many times you roll a pair of fair dice, the chance of coming up 7 or 11 is always the same. Always.

Week Forty-Five

SATURDAY | LOGIC

In Mr. Gardner's fifth grade class, someone accidentally left the hamster cage open, and Fluffy got out. Five kids were in the room, but they all have different stories.

Hannah said:
It wasn't Addison. It was Grace.

Grace said:
It wasn't Graham. It wasn't Addison.

Graham said:
It was Addison. It wasn't Hannah.

Emily said:
It was Graham. It was Grace.

Addison said:
It was Emily. It wasn't Hannah.

Mr. Gardner knows that each student told exactly one lie. Who left the hamster cage open?

Ready for this? Graham did it. Take a look at Hannah's statements. Assume that the second statement is true. If Grace did it, then Addison did it. Two people didn't leave the hamster cage open, so neither Grace nor Addison did it. Now look at Graham's statements. Assume that the first statement is true. Then Addison did it and Hannah did it. Again, there's a contradiction. Besides, Addison has already been eliminated as the culprit. Add Hannah to the innocent list. Take a look at Addison's statements. Assume that the first is true. That means the second is true as well. A third contradiction! And Hannah has already been eliminated. Emily is innocent and the only person who hasn't been eliminated is Graham. He must have left the cage open.

SUNDAY | **GRAB BAG**

Ancient mathematician Diophantus is known as the father of algebra. Not much is known about him, except for this riddle: His youth lasted ⅙ of his life. His first beard lasted $\frac{1}{12}$ of his life. He married after the next ½ of his life. Five years later, his son was born. His son lived ½ of his life. Diophantus died 4 years after his son died. How old was the mathematician?

The first thing to notice is that none of the time spans overlap, and there are no gaps between them. So basically if you knew how long each of these fractions were in years, you could add them to find the old guy's age.

You do have the next best thing: the fractions of his age. With those, you can write an algebraic equation: $\frac{1}{6}x + \frac{1}{12}x + \frac{1}{7}x + 5 + \frac{1}{2}x + 4 = x$. Then you can solve for x.

It's probably easiest to add the 5 and 4 first: $\frac{1}{6}x + \frac{1}{12}x + \frac{1}{7}x + \frac{1}{2}x + 9 = x$. Now find common denominators. You can make this easier on yourself, if you look carefully at the denominators. Notice that 6, 12, and 2 have the same least common multiple (LCM): 12. If you work with those fractions first, you can add them in one fell swoop: $\frac{2}{12}x + \frac{1}{12}x + \frac{1}{7}x + \frac{6}{12}x + 9 = x \rightarrow \frac{1}{7}x + \frac{9}{12}x + 9 = x$. Now find the LCM of 7 and 12. Turns out it's 7 × 12 or 84: $\frac{12}{84}x + \frac{63}{84}x + 9 = x$. Add the fractions: $\frac{75}{84}x + 9 = x$. Subtract $\frac{75}{84}x$ from both sides of the equation: $9 = \frac{9}{84}x$. Multiply each side of the equation by the reciprocal of the fraction to isolate x: $84 = x$.

Week Forty-Six

MONDAY | NUMBER SENSE

The Least Common Multiple (LCM) of two numbers is the smallest number that both numbers will divide into evenly. What is the LCM of 18 and 60?

You can certainly start with listing a bunch of multiples for 18 and a bunch of multiples for 60. That will definitely get you to the right answer: 180. But there is another way.

If you list the prime factors of each number, you can find the LCM quickly. What's a *prime factor*? It's a prime number that divides evenly into another number. (A prime number is a number that has *no* factors. Examples include 2, 3, 5, 7, 11, and so on.) The prime factors of 18 are 2, 3, and 3. That's because $2 \times 3 \times 3 = 18$. The prime factors of 60 are 2, 2, 3, and 5, because $2 \times 2 \times 3 \times 5 = 60$.

But how do you get from these lists of prime factors to the LCM? Look at each list and determine the greatest number of times a factor appears. For example, 2 appears in both lists, but it appears once in the first and twice in the second. So the greatest number of times it appears is twice. The greatest number of times 3 appears in the lists is twice, but 5 only appears in a list once. Now, multiply each factor by itself—the same number of times it appears most often: 2×2, 3×3, and 5. (Five only appears once, so no multiplication is necessary.) Then multiply the results: $4 \times 9 \times 5 = 180$. That's the LCM.

TUESDAY | ALGEBRA

Lines can be written as equations. These equations are created using the slope of a line (how much it slants) and a point on the line. The point-slope form of a line is: $y - y_1 = m(x - x_1)$. The point (x_1, y_1) is on the line, and m is the slope of the line. In another case, the point on the line is the y-intercept—or the point where the line crosses the y-axis. The y-intercept form is: $y = mx + b$. In this case, b is the y-intercept and once again m is the slope. If a line has a slope of $m = 2$, and a point on the line is (1, 1), what is the equation of the line in point-slope form? Then change the equation so that it is in y-intercept form.

There's a lot going on here, but if you take things step by step, you can figure it out. First, substitute 2 for m and (1, 1) for (x_1, y_1) in the point-slope form: $y - 1 = 2(x - 1)$. That's the equation in point-slope form. (Not so bad, eh?)

Now distribute and combine like terms to change the equation into y-intercept form: $y - 1 = 2x - 2 \rightarrow y = 2x - 1$. And now the equation is in y-intercept form.

Week Forty-Six

WEDNESDAY | GEOMETRY

The point (-2, -3) is reflected over the *x*-axis.
What are the coordinates of the point?

A geometric reflection is pretty much the same as a mirror reflection. But since you're only reflecting a point, you don't need to worry about a mirror image—which can be confusing.

When you reflect a point over a line, the new point will be exactly the same distance from the line as the original point. In this case, the original point is in the lower left quadrant of the coordinate plane. When you reflect the point over the *x*-axis, the new point will be in the upper left quadrant. The *x*-value will stay the same, and the sign of the *y*-value will switch. That means the coordinates are (-2, 3).

Of course, if you want to graph the point to see for sure, go for it.

Week Forty-Six

THURSDAY | APPLICATION

The pattern for a knitted afghan calls for 12 skeins of yarn. This produces a blanket that is 50 inches × 70 inches. You want to make that pattern in a smaller size: 30 inches × 36 inches.

How many skeins of yarn do you need?

What you're looking for here is a proportion. But there are so many numbers! Perhaps the easiest approach is to find the area of each afghan. The larger afghan has an area of 50 inches × 70 inches or 3,500 square inches. The smaller afghan has an area of 30 inches × 36 inches or 1,080 square inches.

Now set up a proportion with the areas in one ratio and the number of skeins in the other: $1,080/3,500 = x/12$. Notice how the 1,080 is with the x, and the 3,500 is with the 12? You could set this up in another way. You'll get the same answer as long as you keep these pairs of numbers in matching places, like this: $x/1,080 = 12/3,500$.

Whichever proportion you use, cross multiply and solve for x: $1,080/3,500 = x/12$ ➔ $12,960 = 3,500x$ ➔ $3.7 \approx x$. You can't purchase 3.7 skeins, so 4 skeins it is.

(Try out the other proportion. It's amazing how you get the same answer.)

Week Forty-Six

FRIDAY | PROBABILITY & STATISTICS

From a regular deck of cards (no jokers), you pull one card. What is the probability that it is less than 3 or greater than 10?

First, think about the number of cards that have values less than 3 or values greater than 10. In this deck, ones and twos are in the first category, while jacks, queens, and kings are in the second category. What about the aces? It doesn't really matter which category you put them in, so say aces represent ones.

But remember, there are 4 suits. Therefore, there are 4 aces, 4 ones, and 4 twos, as well as 4 jacks, 4 queens, and 4 kings. That means there are 12 cards with values less than 3 and 12 cards with values greater than 10.

What is the probability that you'll draw a card smaller than a 3? There are 52 cards in the deck, and 12 are smaller than a 3. So that probability is $^{12}/_{52}$ or $^{3}/_{13}$. There are also 12 cards larger than a 10, which means that probability is also $^{3}/_{13}$. The probability of getting a card that's smaller than 3 or larger than 10 is the sum of these probabilities: $^{3}/_{13} + ^{3}/_{13} = ^{6}/_{13}$.

SATURDAY | LOGIC

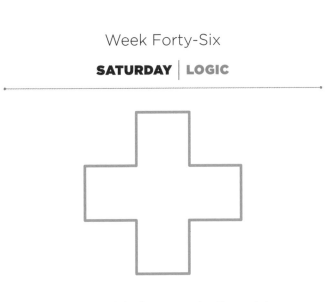

With two straight lines, cut the figure into as many pieces as possible.

Notice that the pieces do not need to be the same size or shape. Trial and error may be the best way to go about this problem. Start with two horizontal or vertical lines. These will only cut the figure into 3 pieces. If you use one horizontal line and one vertical line, you get 3 or 4 pieces—depending where the two lines meet.

Slanted lines that meet in the center of the figure only give 4 pieces as well. But what if you shift these lines up, down, or to the right or left? If done at the correct angle, you can get as many as 6 pieces. For example, if you choose a point near the top of the figure, and draw two slanted lines through that point, you can get 3 parts at the top of the figure, 1 large part at the center and bottom, and 2 parts on the left and right arms of the figure.

SUNDAY | GRAB BAG

Which is larger, the sum of all of the even integers from 1 to 100 or the sum of all the odd integers from 1 to 100?

Y̲ou could add up all of the numbers, but that would take a long, long while. Instead, look at part of the list and see if you can draw any conclusions. Here are the lists of the first five odd numbers and the first five even numbers:

1, 3, 5, 7, 9

2, 4, 6, 8, 10

If you look at the numbers as pairs—one from the top list and one from the bottom list—you can see that in each pair, the even number is larger.

And that affects the sums: $1 + 3 + 5 + 7 + 9 = 25$ and $2 + 4 + 6 + 8 + 10 = 30$.

So it's a good guess that the sum of the even numbers will be larger.

Want to know the actual sums? The evens add up to 2,550, and the odds add up to 2,500.

Week Forty-Seven

MONDAY | **NUMBER SENSE**

What is the percent change from 5 to 20?

Ugh. These give so many of us trouble. You don't need to remember a formula to solve percent change problems. Just think carefully about what is happening in the problem. You're being asked for the percentage that represents the growth from 5 to 20.

First off, this is an increase, so your answer had better be positive.

Next, identify the original value and the new value. In this case, the original value is 5—that's what you start with. The new value is 20—that's what the 5 grew to. How can you describe the change in those values? Subtract the original value from the new value: $20 - 5 = 15$. Now you want to know what percentage this change is, so divide by the original value: $15 \div 5 = 3$.

Next comes the tricky part. The answer you just got is *not* 3%. Nope. You've got to move the decimal point two places to the right to get 300%. How can you tell? Think about the original numbers. Does it make sense to say that the change was only 3%? Not when 5 grew to 20.

Week Forty-Seven

TUESDAY | ALGEBRA

Here are two ways that lines can be written as equations:

point-slope form: $y - y_1 = m(x - x_1)$
y-intercept form: $y = mx + b$

In these equations, m is the slope (the slant of the line), (x_1, y_1) is a point on the line, and b is the y-intercept (where the line crosses the y-axis).

If a line has a slope of $m = -8$, and the line goes through the origin, what is the equation of the line in point-slope form? Then change the equation so that it is in y-intercept form.

Start with the point-slope form. You know the slope is -8, but what is the point on the line? The line goes through the origin, which is (0, 0). And that's the point on the line. Substitute to get $y - 0 = -8(x - 0)$. Simplify to get: $y = -8x$.

In this case, the equation is already in y-intercept form. So what is the y-intercept? You knew from the start: it's 0, because the line goes through the point (0, 0).

Divide this figure into 5 congruent pieces.

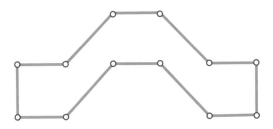

Remember, when figures are congruent, they have the same size and the same shape. You might be able to use straight lines to divide this figure into 5 shapes with the same area—but they won't be congruent.

The answer is actually very simple. Just divide the far left vertical side into 5 equal parts. Do the same on the far right, vertical side. Connect these with jagged lines that mirror the overall shape of the figure. You'll have 5 identical figures that stack on top of one another to make the original figure.

THURSDAY | APPLICATION

A landlord charges 1% of the monthly rent for each day the rent is late. If a rent payment is $750 per month and is 5 days late, how much does the renter need to pay?

Notice that the question is asking for the total—not just the cost of the late fee. So you'll need to find the late fee and then add it to the monthly rent. Start by finding 1% by $750: 0.01 × 750 = $7.50. (You might have done that in your head, but either way, be careful with your place value. It's really easy to make miscalculation and end up with $75.)

Now this is the fee per day, so you'll need to multiply that by 5, which gives the total late fee: $7.50 × 5 = $37.50. Now add this amount to the rent: $37.50 + $750 = $787.50.

Could you approach this differently? Well, of course! For example, you can multiply the percentage by the number of days and then multiply by the rent: 0.01 × 5 = 0.05 → 0.05 × $750 = $37.50.

Finally, you can add.

Week Forty-Seven

To win the annual Smithville Founders Day raffle, the winner must be present. Six hundred tickets were sold this year, and you purchased 20 of them. (The prize is a weekend stay at the Smithville B&B, plus dinner at the only restaurant in town that uses cloth napkins!) You're there for the drawing, watching as 10 tickets are drawn—and none of the ticket holders are there. What is your chance of winning on the eleventh draw?

The previous 10 tickets are not put back. This changes the number of total outcomes. There were 600 tickets, but now there are 590 tickets to choose from. But your 20 tickets are still in there, giving you a $^{20}/_{590}$ or $^{2}/_{59}$ chance of winning on that 11th draw.

Obviously, the odds are not in your favor. But if they draw a few more tickets and none of those ticket holders are present, your chances go up. By how much is another problem altogether.

Week Forty-Seven

SATURDAY | LOGIC

Harold has 4 scrap lengths of chain, each with 3 links. He wants to make a loop with the chain. To do this, he'll need to cut some links, hook links together, and then solder the cut links. What is the minimum number of times he must cut a link and then solder it closed?

Harold could cut 1 link of each of the scrap chains and then solder the 4 pieces together. This would translate to 4 cuts and solders. But there is a quicker way that takes only 3 cuts and solders. Cut all of the links of 1 scrap chain, and use these 3 cut links to connect the remaining 3 scrap chains into a loop.

Week Forty-Seven

SUNDAY | GRAB BAG

What is the pattern in this list of numbers?

6, 19, 58, 175

Finding the patterns in numbers can be pretty difficult. Sometimes trial and error helps, while other times, the pattern will seem to emerge quickly. The goal is to find the consistent rule that leads you from one number to the next. So what do you do to get from 6 to 19? Well, there are many different options. Here are two: 6 + 13 = 19, 25 - 6 = 19. Do either of these work for the next pair of numbers? 19 + 13 ≠ 58 and 25 - 19 ≠ 58. There must be another pattern.

The numbers are getting larger pretty quickly. So could multiplication be involved? There is no number that when multiplied by 6 gives you 19, but 6 × 3 = 18. Add 1 and you have 19. Now test the next pair. 19 × 3 = 57. Add 1 and you've got 58. Looking good! 58 × 3 = 174. Add 1, and you've got 175. So the pattern is to multiply the number by 3 and then add 1.

Week Forty-Eight

MONDAY | **NUMBER SENSE**

What is the percent change from 12 to 8?

Again with the percent change. This time you're going from a larger number to a smaller number. Therefore, you're actually looking for a percent decrease. In terms of the process, this doesn't matter. But your answer should be negative, showing that decrease.

Subtract the original number from the new number and then divide by the original number: $(8 - 12) \div 12 = -4 \div 12 = -0.33$. First off, you've got a negative number, which shows the decrease. You still need to make sure that you're writing the percentage correctly. Move that decimal point two places to the right to get -33%. You can write that as a negative percentage or use words to describe the change: a decrease of 33%.

A horizontal line goes through -2 on the *y*-axis.
What is the equation of this line? Explain why.

You might want to sketch the graph or use your imagination to picture it. Since the line is horizontal, it goes from left to right, with no slant at all. And every point on the line has a *y*-value of -2, while the *x*-values change. This means that every single possible *x*-value is on the line. So the equation of the line is $y = -2$. In fact all horizontal lines have this format, depending on where the line intersects the *y*-axis.

WEDNESDAY | GEOMETRY

A three-dimensional figure is made up of small cubes.
Each dimension of the figure has 5 of these small cubes.
How many cubes are in the figure?

It might help to draw the figure. Because it is made up of cubes and has dimensions of 5 cubes by 5 cubes by 5 cubes, the figure itself is a larger cube. To find the number of smaller cubes that fit inside the larger cube, multiply the dimensions: $5 \times 5 \times 5 = 125$ smaller cubes.

If the multiplication doesn't make sense, you can think of the problem this way: Layer the smaller cubes to make the larger cube. One layer will be 5 cubes wide by 5 cubes long or 25 cubes. There are 5 layers, so there are 25 + 25 + 25 + 25 + 25 or 125 cubes. Multiplication or addition—the answer is the same.

Week Forty-Eight

THURSDAY | APPLICATION

Julian is financing the purchase of a used car. The price of the car is $18,000, and he's going to make a $3,000 deposit. How much will he pay over the life of the loan if the annual interest rate is 7%, compounded monthly over 48 months? Use this formula: $A = P(1 + r/n)^{nt}$. A is the total amount; P is the principal; r is the interest rate per compounding period; and n is the number of compounding periods.

First, Julian needs to identify all of his variables. He doesn't know the total amount he'll pay for the car. (That's A, what he's trying to figure out.) The principal is the amount he borrowed or the price of the car minus the down payment: $18,000 - $3,000 = $15,000.

The interest rate per compounding period requires a little bit of work. Since the interest rate is annual, but the interest is compounded monthly, divide 7% by 12 to get r: $0.07 \div 12 = 0.006$. Finally, the number of compounding periods or n is 48 months.

Substitute, and follow the order of operations:
$A = 15000(1 + 0.006)48 \rightarrow A = 15000(1.006)48 \rightarrow$
$A = 15000(1.33) \rightarrow A = $19,950.

So Julian will pay nearly $5,000 in interest alone.

Week Forty-Eight

FRIDAY | PROBABILITY & STATISTICS

A kid in your neighborhood loves to create board games. His most recent creation has some complicated rules. At each turn, the player must roll a die and flip a coin. What is the probability that a player will roll a 3 *and* flip heads?

Take each rule one by one, starting with the die. A die has 6 sides, so there are 6 possible outcomes. Only one of those outcomes can be a 3, which means that the probability of rolling a 3 is $\frac{1}{6}$. A coin has 2 sides, so there are 2 possible outcomes. Only one of those outcomes can be heads, so the probability of getting heads is $\frac{1}{2}$.

Now comes the hard part. How do you know the probability of rolling a 3 and flipping heads? Unlike other word problems—when *and* means addition—in probability, *and* means multiplication. So you can multiply the two probabilities to get $\frac{1}{12}$.

Week Forty-Eight

SATURDAY | LOGIC

Three people check into a hotel. The desk clerk tells them that each room is $10. The three people pay $30 and go up to their rooms. The hotel manager comes back from break, and tells the desk clerk that he should have charged $25 for three rooms. He asks for $5 to take up to the guests. On the way up, he realizes that it would be easier to give each person $1 or $3 total. He pockets the $2. In the end, each person paid $10 and got a $1 refund. So in total, they paid $27. The manager kept $2, which brings the total to $29.

What happened to the missing $1?

Pay close attention to how you're multiplying, adding, and subtracting. The manager refunded $3, which means the room fee plus the amount the manager kept was $27. The true cost of the three rooms was $25. The guests paid an extra $5, $2 of which the manager kept. That means the hotel and the manager have a total of $27. The manager returned $3, and $27 + $3 = $30. So, all of the money is accounted for. In other words, there is no missing dollar!

Week Forty-Eight

SUNDAY | GRAB BAG

How can you arrange 6 matchsticks to create
4 identical triangles?

More than likely this problem sounds pretty familiar. But do you remember how it's done? One approach is to grab a box of matches—or toothpicks or pencils—and try some things out. There are several different answers, but here's the most common: Make a square out of 4 of the matchsticks. Then use the remaining 2 matchsticks to create the diagonals of the square. The 4 triangles formed by the square and its diagonals are identical. (Now, this isn't an *exact* answer, because of course if all of the matchsticks are the same length, they won't be long enough to form actual diagonals. But you get the drift anyway.)

Another approach is to create a triangle using 3 of the matchsticks. Then use the remaining 3 matchsticks to create an inverted triangle *inside* the first triangle. This creates 4 identical triangles, but this time, the matchsticks that form the inside (smaller) triangle will overlap the matchsticks that form the outside (larger) triangle.

And one last solution, which is much more exact: you can create a three-dimensional figure, called a *tetrahedron*. This figure is a pyramid: it has 4 triangular sides (or faces), with 6 edges (or matchsticks). The 4 triangles are identical.

Week Forty-Nine

MONDAY | NUMBER SENSE

What is $-8(4 - 6)^3$?

You're really going to have to rely on the order of operations here. Start with the parentheses. Because you're subtracting a larger number from a smaller number, you'll find the difference and make the answer negative: $-8(-2)^3$. Next take care of that exponent. You're cubing a negative number, so the result will be negative: $-8(-8)$. Finally multiply the two negative numbers to get a positive number: 64.

What would happen if you did the process out of order? Well, you'd get a different answer, of course. And since arithmetic needs precision, the rules matter.

TUESDAY | ALGEBRA

A vertical line goes through 12 on the *x*-axis.
What is the equation of that line? Explain why.

A sketch of the graph could be a good idea here. The line is vertical, so that means it goes straight up and down with no slant. All of the points on the line have an *x*-value of 12. But the *y*-values can be anything at all. So the equation is $x = 12$. In fact, all vertical lines have this format, depending on where the line intersects the *x*-axis.

Week Forty-Nine

WEDNESDAY | GEOMETRY

Move 4 matchsticks in the figure below to create 6 identical triangles.

There are 12 matchsticks in all, and you need 6 triangles. Therefore, the triangles must share sides. (You can only make 4 triangles that share no sides. That's because 3 × 4 = 12.) The solution? Make a 6-sided polygon, a hexagon. Do that by removing the top 2 and bottom 2 matchsticks. Use 2 of these 4 matchsticks to close off the bottom and the top of the figure. This creates the hexagon. But you don't have 6 identical triangles yet. Take the 2 remaining matchsticks and place them horizontally so that their ends meet where the 4 matchsticks in the middle meet. In this way, you've created the diagonals of the hexagon, which creates 6 identical triangles.

Having trouble understanding? Give it a whirl with real matchsticks.

Week Forty-Nine

THURSDAY | **APPLICATION**

Jordan has decided to live life more modestly. She has trimmed her workout wardrobe down to 5 tops, 3 pairs of pants, and 2 pairs of shoes. How many different outfits does she have?

There aren't many articles of clothing, so pair each top with a pair of pants, and then each top and pant combination with a pair of shoes. Jordan has 5 tops for every pair of pants. That's 5 + 5 + 5 or 5 × 3 or 15. But now you've got to address the shoes. There are 15 top and pant options and 2 pairs of shoes: 15 + 15 or 15 × 2 or 30. That means Jordan has 30 outfits to choose from. That's a lot of combinations from such a small closet.

FRIDAY | PROBABILITY & STATISTICS

10TH Percentile	$12,276	60TH Percentile	$68,212
20TH Percentile	$21,432	80TH Percentile	$112,262
40TH Percentile	$41,186	90TH Percentile	$157,479
50TH Percentile	$53,657	95TH Percentile	$206,568

The table above shows the household incomes in the U.S. for a recent year. This data is divided into percentiles, which give the percentage of people who fall at or below the given income level. (The data comes from the U.S. Census, and does not include the 30th or 70th percentiles.) What does this data say about income in the U.S.? If the population is 318 million, how many people earn more than $206,568 per year? How many people earn less than $12,276?

Based on the data above, lower incomes are generally clustered together. The data is more spread out when looking at percentiles greater then 60%. There are 15.9 million people in the 95th percentile, and 31.8 million people in the 10th percentile. That demonstrates that a small percentage of people have relatively high incomes, while a larger percentage are earning much less.

Since there is no information about the households represented, it's difficult to draw additional conclusions. For example, it could be that families make up the largest number of low earners. More information is needed to make that distinction.

Week Forty-Nine

SATURDAY | LOGIC

Your office is hosting a gag-gift exchange. Some gifts are wonderful and other gifts are terrible. It's your turn, and you can choose from three gifts. The gifts are wrapped, so you can't see inside, but you do know this: One of the three gifts is an extra vacation day. Woo-hoo! And you know the other two gifts are real stinkers: a mug that says "World's Worst Employee" and a 2013 calendar. You make your choice, and the boss lets out a secret: one of the remaining gifts is the calendar. He gives you a chance to switch your prize. Should you?

Take a look at the chances of getting any of the three options, at any point in the game. You have a 1 in 3 chance of getting the extra vacation day when you make your first choice. Since you know that one of those remaining two gifts is the calendar, you have a 2 in 3 chance that the second remaining gift is the extra vacation day. So, yep, you should definitely switch.

Week Forty-Nine

SUNDAY | GRAB BAG

How can you arrange four 9s to make 100?

Yuo've been given no other instructions, which means you can put these four 9s together in any way you wish. You can use three of the 9s to make 999, or you can use two of the 9s to make 99. Another option is to use mathematical operations between the 9s. For example, you could just add them: 9 + 9 + 9 + 9. But that gives you 36, not 100. What if you multiply and add? 9 × 9 + 9 + 9 = 99. Not quite!

In fact, you're going to have to get fancy. Try starting with 99. If you add 1, you'll get 100. And since 9 ÷ 9 = 1, you can add 9 ÷ 9. So the correct answer is 99 + 9 ÷ 9. Notice that if you don't apply the order of operations correctly, you can't get the correct answer!

Week Fifty

What is 6 – 10 + -3?

There are a few things going on here. First, you are asked to subtract a larger number from a smaller number. Next, you are asked to add a negative number. Because the problem requires only subtraction and addition, you can simply solve from left to right. So start with 6 – 10. To do this, ignore the signs, find the difference, and make the answer negative: -4. Next add -3, which means adding a negative and a negative. To do this, ignore the signs, add, and make the answer negative: -4 + -3 = -7. And that's all there is to it.

Week Fifty

TUESDAY | **ALGEBRA**

The midpoint between two points on a coordinate plane can be found using the midpoint formula:

$$\left(\frac{x_1 - x_2}{2}, \frac{y_1 - y_2}{2} \right)$$

Use the midpoint formula to find the midpoint between (3, -8) and (-12, 4).

To get your grounding, picture the coordinate plane. The first point could be called (x_1, y_1), and the second point could be called (x_2, y_2). (Honestly, if you switch the two, it's no biggie. Just be sure that you keep the two x-values in the same place in the formula, and the two y-values in the same place in the formula.)

Now you can simply substitute and solve. Start with the numerators. Here's the numerator of the x-value: 3 – -12 = 3 + 12 = 15. Here's the numerator of the y-value: -8 – 4 = -12. Now divide each of these by 2 to get the midpoint: ($^{15}/_2$, -6).

So why is each value divided by 2? The midpoint is halfway between each of the points, so you're multiplying by $^1/_2$ or dividing by 2.

WEDNESDAY | GEOMETRY

Divide the shape below into 4 identical pieces that do not overlap.

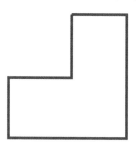

Think of this figure as a backwards *L*. Naturally, you'll start by trying to create squares. But with 2 line segments, drawn from the "crook" of the backwards *L*, you only get 3 squares. What about triangles? You can get 4 triangles by drawing 2 line segments: 1 from the bottom left vertex to the top right vertex, and the other from the bottom right vertex to the vertex at the "crook" of the backwards *L*. But the triangles at the bottom right are larger than the triangle at the bottom left and the triangle at the top left.

Actually, though, the answer is rather simple. All you need to do is replicate the backwards *L* 4 times, so that they fit inside the larger figure. One of these fits inside the "crook" of the backwards *L*. Then you can draw 2 line segments from the bottom side of the smaller *L* and from the right side of the smaller *L*. That creates 4 backwards *L*s, 1 flipped upside down at the top, 1 at the "crook," 1 at the bottom right, and 1 flipped and at the left side.

Week Fifty

Your total at the diner is $10.50. The tax is 7%, and you want to give a 20% tip. You give your server $20 (which includes the tip). How much change should you get back?

First, you need to figure out the tax. Multiply $10.50 by 7%: $10.50 × 0.07 = 0.74. So you'll need to add 74¢ to your total: $10.50 + 0.74 = $11.24. Now, find the 20% tip: $11.24 × 0.2 = $2.24. Add that to the previous total: $11.24 + $2.24 = $13.48. Finally, subtract this amount from $20 to find out your change: $20 – $13.48 = $6.52. Cheap date!

FRIDAY | **PROBABILITY & STATISTICS**

A state issues license plates using the following pattern: 6 characters total, with 3 letters followed by 3 numbers. What is the greatest number of registered cars in that state?

Each license plate must be unique. Therefore, the combination of letters and numbers must not repeat. But there is no restriction on whether letters or numbers can repeat in a given license plate. In other words, it's fair game to have a plate LLL-333.

This thinking can help you find the number of possible license plates, which will give the most number of registered cars in the state. What are all of the possibilities for the first character? Since there are 26 letters in the alphabet, that would be 26. The second character has the same number of possibilities, since repeating characters is A-OK. Same with the third character.

But when you get to the fourth character, things change a bit. There are 10 possible numbers: 0, 1, 2, 3, 4, 5, 6, 7, 8, and 9. The same is true for the last two characters.

To find the total possible license plate numbers, multiply each of these values:
$26 \times 26 \times 26 \times 10 \times 10 \times 10 = 17{,}576{,}000$. That's a lot of cars.

You have 8 gold dollar coins, but only 1 is real. The real coin weighs more than each of the counterfeit coins. What is the minimum number of times you can weigh the coins, using a balancing scale, to find out which coin is the real one?

Put 3 coins on each side of the scale and leave the remaining 2 coins on the table. One of two things will happen. In the first scenario, the scales balance. This means that 1 of the 2 coins on the table is the real coin. To find out which one is real, remove the coins on the scale, and place 1 coin from the table on each of the 2 trays of the scale. The coin on the lower tray is the real one.

In the second scenario, the 6 coins on each tray of the scale do not balance. Remove the coins from the lighter side of the scale. Then take 1 coin from the heavier side and put it on the table. Put 1 of the 2 remaining coins on each side of the scale. If the scale balances, the real coin is on the table. If it doesn't, the real coin will be on the heavier side of the scale.

SUNDAY | **GRAB BAG**

What number between 1 and 10 would make the following true?

$$\frac{x}{x} - \frac{x}{2} + \frac{x}{4} = \frac{x}{12}$$

You can certainly use trial and error to find the answer, but what would happen if you got a common denominator for each of the fractions with numbers in the denominator? The common denominator is clearly 12, which makes the problem look like this:

$$\frac{x}{x} - \frac{6x}{12} + \frac{3x}{12} = \frac{x}{12}$$

Now combine like terms and simplify:

$$\frac{x}{x} - \frac{3x}{12} = \frac{x}{12}$$

$$\frac{x}{x} = \frac{4x}{12}$$

What value would make this equation true? The missing value must be 3. Check the answer by substituting into the original equation.

Week Fifty-One

MONDAY | NUMBER SENSE

Which number is the largest?

$$412 \quad 4^{12} \quad 2^{14} \quad 41^2$$

(Try without using a calculator and then check your answer.)

The easy way to solve this problem is to use a calculator to find the values of the last three numbers. But why do all of that work? (Besides, how is that really working your gray matter?)

Look at the first and last value to begin with. Which one is larger? 41^2 is pretty close to 40^2, which is 1600. So clearly, 412 is less than 41^2.

And based on what you know about exponents—mainly that they grow really quickly—it's a good bet that 4^{12} and 2^{14} are going to be bigger than both 412 and 41^2.

Take a guess at which one is larger. 4^{12} has a larger base than 2^{14} but a smaller exponent. (The base is the big numeral, and the exponent is the small numeral.) And because 2 is half of 4, it's a good guess that 4^{12} is the larger of the two middle values.

The calculator tells you that 4^{12} is about 16.8 million, while 2^{14} is 16,384. So the answer is clear: 4^{12} is the largest number, followed by 2^{14}, 41^2, and 412.

You didn't even have to use a calculator to make an educated guess.

The distance formula is used to find the distance between two points that are graphed on a coordinate plane. It looks like this:

$$d = \sqrt{(x_2 - x_1)^2 + (y_2 - y_1)^2}$$

What is the distance between (0, 0) and (6, 6)?

That's a really scary formula for some folks. And if you're one of those folks, remember to take things step by step. First, identify which variable is which in the points given.

To tell you the truth, it doesn't matter which point is (x_1, y_1) and which is (x_2, y_2), just as long as you're consistent. Why not just go in order? Start by dealing with the stuff under the square root symbol:
$(0 - 6)^2 + (0 - 6)^2 = 6^2 + 6^2 = 36 + 36 = 72$.
Now you can take the square root: $\sqrt{72}$.

You can use a calculator to estimate the value of this square root, or you can use some more advanced algebraic tricks to simplify the radical:
$\sqrt{72} = \sqrt{36} \times 2 = 6\sqrt{2}$.

Week Fifty-One

WEDNESDAY | GEOMETRY

What is the smallest number of matchsticks you need to make 5 equilateral triangles?

It's probably not as many as you think. Turns out 9 is plenty. Make 1 triangle using 3 of the matchsticks. Then build on—adding 2 matchsticks to the right of this triangle—to make a second triangle that points in the opposite direction. (If the first triangle pointed up, the second points down, for example.) Build on again to the right—adding 2 matchsticks to make a third triangle. Finally use the last 2 matchsticks to make the fourth triangle above the first three triangles. But what about the fifth triangle? That's the larger triangle formed by all of the smaller ones. You can also work this problem backwards. Use 6 of the matchsticks to make one triangle. (Each side of the triangle uses 2 matchsticks.) Then form a smaller triangle inside this one, using the remaining 3 matchsticks.

THURSDAY | APPLICATION

You need to create an 8-character password. You must use letters, numbers, and symbols. The letters you want to use are the first letters of your children's names: R, J, and P. The numbers are the numbers in your anniversary: 7, 2, 3. And the symbols you have chosen are & and $. How many unique passwords can you create? (You cannot use any of the letters, numbers, or symbols more than once.)

Think of this in terms of finding the possibilities for each of the 8 characters in the password. There are 8 possibilities in all. If you're choosing the characters one by one, what are all of the possibilities of the first character? Since you haven't assigned any characters yet, that would be 8. Now, for the second character, you only have 7 possibilities left—because you used one possibility for the first character. Seeing a pattern here?

As you go along, you'll have one fewer possibility to use. Multiply each of these possibilities to find the total: $8 \times 7 \times 6 \times 5 \times 4 \times 3 \times 2 \times 1 = 40{,}320$.

Wonder how long it would take a computer to hack into that password.

Week Fifty-One

FRIDAY | PROBABILITY & STATISTICS

Social Security numbers are in this form: 000-00-0000. These numbers are given to each citizen of the United States. When someone dies or renounces their citizenship, their number goes back into the pool of numbers that can be assigned to infants and other new citizens. The numbers 0 through 9 can be used in any place and can be repeated. How many possible Social Security numbers are there?

Because the numbers can be repeated, there are 10 possible numerals for each place. And there are 9 places. That means you'll multiply 10 by itself 9 times, which is the same thing as 10^9. That's 1 with 9 zeros behind it, or 1 billion. No chance of running out of Social Security numbers anytime soon.

Week Fifty-One

SATURDAY | LOGIC

Mazzy and Carly are identical twin sisters. One always lies and the other always tells the truth. You ask one of them, "Does Mazzy always lie?" She answers, "Yes." Were you speaking to Mazzy or Carly?

This is a common spin on a common logic problem. Clearly there are two possible situations: either you have spoken to Mazzy or you have spoken to Carly.

Assume you were speaking to Carly. If she is the truth teller, then her answer, "Yes," means that Mazzy is the liar. That works, since one is a liar and one is not a liar.

Now assume you were speaking to Mazzy. If she is the truth teller, then her answer should be, "No." She cannot be a truth teller *and* a liar.

This means that you were speaking with Carly.

Week Fifty-One

SUNDAY | GRAB BAG

A train leaves Baltimore, MD, going south toward Richmond, VA. At the same time, another train heads north from Richmond toward Baltimore. The Baltimore train is traveling 105 miles per hour (mph). The train from Richmond is traveling 95 miles per hour (mph). The distance between the two cities is 160 miles. When will the two trains pass one another?

For this problem, you need the distance formula, which is easy to remember if you know what *miles per hour* means. Of course, mph is the speed at which something is traveling. *Per* means division, so you can write *mph* as an equation: $r = d \div t$, where r is the rate or speed, d is the distance in miles, and t is the time in hours. But you may remember the distance formula this way: $d = rt$.

In this problem, you know the distance between the starting points of the two trains. And you know how fast each train is going. In fact, for each train, you can write the distance as the product of the rate and time. For the Baltimore train, that would be $105t$. For the Richmond train, the distance can be written as $95t$.

Here's the tricky part. You can now write this equation: $105t + 95t = 160$. In other words, the sum of the distances each train has travelled when they pass is equal to the total distance between the two cities. Now you can easily solve for t to find out when the trains pass: $105t + 95t = 160 \rightarrow 200t = 160 \rightarrow t = 0.8$ hours. So the two trains will pass 0.8 hours after they each leave the station.

MONDAY | **NUMBER SENSE**

You need to create a 3-digit, even number. How many of these numbers can be formed using the numerals 3, 4, 5, 6, and 8? (Any of these numerals can be repeated.)

Y ou've got 3 digits and 5 different numerals, and you need create even numbers. The first thing to think about is what it means for a number to be even. It must end in a digit divisible by 2, right? That means there are 3 options for the ones place: 4, 6, and 8. How many options are available for the tens place? Any of the numerals can appear in this place. So there are 5 options. Finally, how many options are available for the hundreds place? That would be 5 again.

To find out the total number of 3-digit numbers, multiply the options for each place: 3 × 5 × 5 = 75. So there are 75 different numbers that can be created. Don't believe it? Write down all of the possible numbers to check.

Week Fifty-Two

TUESDAY | ALGEBRA

What is the fifteenth number in the sequence below?

2, 6, 12, 20, 30, 42, . . .

Write an algebraic expression that describes the pattern.

So you have another sequence here. The best approach is to figure out the rule, but extending the sequence out to the fifteenth number is going to be a lot of work. Luckily there's another option.

First, the rule. You may have noticed that to get each number, you add an even number, starting with 4 and continuing with the next larger even number. So $2 + 4 = 6$, $6 + 6 = 12$, $12 + 8 = 20$, and so on. You can continue this list until you get to the fifteenth number, which is 240. But you don't have to do it that way.

If you think of each number in the sequence as a position, you see that the first number is 2, the second number is 6, the third number is 12, and so on. One common way to describe this is using the variable n. For the first number, $n = 1$ and for the fifteenth number, $n = 15$. This gives you another way to describe the pattern. When $n = 2$, $2(2 + 1) = 6$. When $n = 3$, $3(3 + 1) = 12$. How can you write this as an algebraic expression? $n(n + 1)$. Now you can find the fifteenth number, easy-peasy: $15(15 + 1)$. Check it out. You'll see that you have the same answer as above.

Week Fifty-Two

WEDNESDAY | **GEOMETRY**

Remove 1 matchstick from the ones below. Then rearrange the matchsticks to create 6 identical triangles.

First, count the matchsticks that you have now. There are 13 of them, so when you remove 1 of them, you have 12 in all. Think about that number 12. It's the same as 4 × 3, and there are 3 sides in a triangle. But that only gives 4 triangles, and you need 6. So clearly, some of the triangles will share sides. In fact, all of the triangles will share sides.

But go back to that 12, which also has factors 2 and 6. There are 6 sides in a hexagon. Could the triangles come from that? The answer is yes. Create the hexagon, and then use the remaining 6 matchsticks to create the diagonals. That uses up all of the matchsticks and creates 6 identical triangles.

A stock price decreases by 20%. By what percent must it increase to reach the original price?

Was your first guess 20%? That's a great guess. But unfortunately, it's not the correct answer. See, the stock price has changed, so the percentage changes, too.

The easiest way to approach this problem is by assigning the stock price a value. And since you're finding percentages, why not make the original stock price $100? Twenty percent of $100 is $20. That means the new stock price is $80. You'll need to add $20 to $80 to get back to the original price of $100, and $20 of $80 is what percent? Divide to find out: 20 ÷ 80 = 0.25 or 25%.

So the new (lower) stock price will need to increase by 25% to get back to the original stock price.

FRIDAY | PROBABILITY & STATISTICS

The margin of error is the amount of random sampling error in a survey result. In other words, it's how far the result is expected to be from the sample if the entire population were questioned. This is the general formula for the margin of error:

$$z \times \sqrt{\frac{\hat{p}(1-\hat{p})}{n}}$$

The z stands for the z-value, which is taken from a table. That p with a funny-looking hat is the sample proportion, which is the percentage of the sample that responded a certain way. And the n is the sample size.

You are conducting a poll for a mayoral candidate. You want a 95% confidence level (which has a 1.96 z-value). Sixty percent of the respondents said they will vote for your candidate, and the sample was 2,000 people. What is the margin of error?

Because you know the formula and everything that goes into it, you merely need to plug the values in and evaluate. Start with the sample proportion, which is 60%. Turn that into a decimal and plug it into the numerator of the formula: 0.6(1 - 0.6) = 0.6 × 0.4 = 0.24. Now divide by the sample: 0.24 ÷ 2000 = 0.00012. Take the square root of this to get ±0.011. Now multiply by 1.96, and you'll get ±0.02 as the margin of error. That means you can expect that between 58% and 62% of the voters will vote for your candidate.

A new family with 3 children moves into the neighborhood. The next-door neighbor visits and says to the mother, "How old are your children?" The mother says, "The product of their ages is 36, their ages each have one digit, and the sum of their ages is the same as your house number." The neighbor thinks for a moment and then says, "I don't have enough information." The mother tells the neighbor that her oldest child is playing in the backyard, and then the neighbor knows how old the 3 children are.

What is the house number? How old are the children?

The product of the children's ages is the real key here. What 3 numbers can you multiply to produce 36? There are 4 sets: $3 \times 3 \times 4$, $2 \times 2 \times 9$, $2 \times 3 \times 6$, and $6 \times 6 \times 1$. Now find the sums of these 4 sets of numbers: $3 + 3 + 4 = 10$, $2 + 2 + 9 = 13$, $2 + 3 + 6 = 11$, and $6 + 6 + 1 = 13$.

The next clue is that the neighbor didn't have enough information. Since the neighbor knows her own house number, the problem must be that 2 options have a sum equal to her house number.

Those are $2 + 2 + 9 = 13$ and $6 + 6 + 1 = 13$. So now you know that the neighbor's house number is 13.

Since the oldest child is playing outside, the neighbor knows that 2 of the children are 2 years old and the oldest is 9 years old. In the other option, the twins are older, meaning that the oldest cannot be playing outside.

Week Fifty-Two

SUNDAY | GRAB BAG

Fibonacci has 2 rabbits—1 male and 1 female. Once they are 2 months old, they are old enough to mate and have a pair of rabbits. The next month, the rabbits that are old enough to mate have 1 more pair of rabbits. This continues, with each pair of rabbits bearing 1 pair of rabbits each month, for 12 months.

How many *pairs of rabbits* does Fibonacci have at the end of the year?

This famous math problem helps illustrate the importance of sequences, or lists of numbers. Of course, there are many different ways to approach this problem, but it might help to start with a list. If you can find the pattern, you can then find the solution to the problem.

The first number in the list is 1, because there is 1 pair of rabbits to begin with. In the second month, this pair of rabbits is still not old enough to have babies, so there is still 1 pair of rabbits. In month 3, the pair of rabbits bears 1 pair of rabbits, so there are 2 pairs of rabbits. We're in the fourth month now, when there is 1 pair of rabbits old enough to mate. So there are 3 pairs of rabbits. In the fifth month, there are 2 pairs of rabbits old enough to mate, so that gives 2 pairs of newborns plus the 3 pairs of rabbits, or 5 pairs of rabbits in total. And in the sixth month, there are 3 pairs of rabbits old enough to mate, who have 3 pairs of newborns. Add the 5 pairs, and you get 8 pairs of rabbits.

See the pattern yet? The clue was in the problem itself. Fibonacci discovered this famous pattern: 1, 1, 2, 3, 5, 8, 13, 21. . . To generate this sequence is simple. Just add the 2 previous numbers in the sequence to get the next number. Keep going to the twelfth place, and you'll find that Fibonacci has 144 rabbits by the end of the year.

Week Fifty-Three

MONDAY | NUMBER SENSE

The natural starting point is to add 7 + 7 + 7 + 7 + 7 + 7 + 7, but that only gets you to 49.

You need to combine some of the 7s as numerals in three- or two-digit numbers. (You won't have a four-digit number, because the answer is only three digits.) To get to the answer more quickly, take a look at the last digit of 868. You need to add 7s so that the last digit is 8.

That means you'll have four numbers that you're adding (7 + 7 + 7 + 7 = 28). Start out with 77 + 77 + 77 + 7 = 238. Nope. What about 777 + 77 + 7 + 7? The answer is 868, so that's it.

Week Fifty-Three

TUESDAY | ALGEBRA

What is the 100th number in the following sequence?

4, 7, 12, 19, 28, . . .

Warning! This is a toughie. Your first instinct might be to find the difference between the numbers: 7 – 4 = 3, 12 – 7 = 5, 19 – 12 = 7, 28 – 19 = 9. There is a pattern here, but it'll be awfully difficult to find the 100th number.

Instead, it's a good idea to find the algebraic expression that works to get each number. And the big clue here is to think of the *position* of each number, not the number itself. If that position is called n, you can say that when n = 1, the number is 4, and when n = 2, the number is 7. So how can you write an expression with n? Take a look at what happens when n = 2. What can you do to 2 to get 7? You can just add 5, but that doesn't work with the other numbers in the sequence.

You're going to have to get fancy here. Take a look at the n values. When n = 2, the number in the sequence is 7: 2 x 2 = 4 and 4 + 3 = 7. What about when n = 3? 3 x 2 = 6 and 6 + 3 ≠ 12. So we need to think differently. 2 x 2 = 2^2, so try squaring n: 1^2 = 1 and 1 + 3 = 4. Also, 2^2 = 4 and 4 + 3 = 7. Ta-da!

So check out this rule for another number, say the fourth number in the sequence: 4^2 + 3 = 16 + 3 = 19. Is 19 the fourth number? Yep. So you've found the rule. As an algebraic expression, this looks like n^2 + 3, where n is the *position* of the number in the sequence. Plug in 100 for n and you'll find that 10,003 is the 100th number in the sequence.

ABOUT THE WRITER

A self-proclaimed "math evangelist," Laura Laing has no doubt that everyone can do math—in their own way. Her writing has appeared in a variety of publications, including *Parade*, *Parents*, and Southwest Airlines' *Spirit* magazine. She also creates math curriculums, is the author of two additional math books—*Math for Grownups* and *Math for Writers*—and blogs at www.mathforgrownups.com.

NOTES